CONTENTS.

CHAPTER I.

PAGE

Cupels and Flux........................... 1

CHAPTER II.

Assay of Ores................... 5

CHAPTER III.

Notes on the Assay of Ores........................ 29

CHAPTER IV.

Fluxes and Reagents.............................. 48

CHAPTER V.

Slags............... 55

CHAPTER VI.

The Assay of So-called Refractory Ores 61

CHAPTER VII.

Amalgamation and Chlorination Tests............... 82

viii   *CONTENTS.*

CHAPTER VIII.

PAGE

The Assay of Ore containing Coarse Metal 88

CHAPTER IX.

The Assay of Copper Ores and Copper-bearing Materials
for Gold and Silver............................. 91

CHAPTER X.

The Dry Assay of Mercury......................... 100

CHAPTER XI.

Volumetric Determination of Copper............... 108

CHAPTER XII.

Volumetric Determination of Lead................. 130

APPENDIX A.

Supplementary Notes on the Assay of Ores........... 134

APPENDIX B.

Atomic Weights and Tables 141

A MANUAL OF ASSAYING.

CHAPTER I.

CUPELS AND FLUX.

1. To Make Cupels.

Take a miner's pan (Fig. 1) full of bone ashes and pass them through a 40-mesh sieve, moisten sufficiently to make them cohere

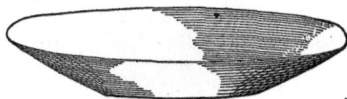

FIG. 1.

slightly, like flour when compressed in the hand (about 1 oz. of water to a pound of bone ashes). Work them between the hands until the moisture is evenly distributed.

Place a 1½-inch cupel mould (Fig. 2) on a
smooth iron plate or anvil, fill it with the
moistened bone ashes, and drive the die down
by a number of light blows with a mallet
(Fig. 3), turning the die after each blow.

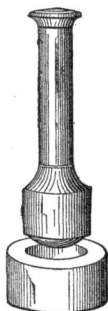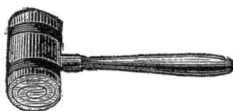

Fig. 2.　　　　　　　Fig. 3.

Move the mould on the iron plate so as to
give the cupel an even base, then turn the
mould and die upside down, and push the
cupel out with the die. Turn cupel when
taking it off. Place the cupels on a dry
board, and dry them slowly and carefully in
a warm room.

Too much water makes the cupels less porous, and if too little water is used, they will crumble. If the cupels stick in the mould, the bone ashes are too dry or they are compressed too hard.

2. Flux for Gold Ore having a Silicious (Quartz) Gangue.

Litharge, PbO................100 parts
Bicarbonate of sodium, $HNaCO_3$..100 "
Wheat flour................. 4 "

Spread a rubber or an oil cloth on the mixing-table. Weigh out 5 lbs. litharge, PbO; 5 lbs. bicarbonate of sodium, $HNaCO_3$; and $3\frac{1}{4}$ ozs. wheat flour, and pass them through a 40-mesh sieve, on the cloth. Spread the mixture out evenly, then, by lifting a corner of the cloth and drawing it forward, roll the mixture on itself, then take another corner in the same way; occasionally spread out the mixture and proceed as before until it is thoroughly mixed. Place the mixture in a box for future use.

2a. Flux for Gold Ore having a Silicious (Quartz) Gangue.

Per cent.
by Weight.

Litharge, PbO........................ 58
Sodium bicarbonate, $NaHCO_3$.......... 30
Potassium carbonate, K_2CO_3. 10
Wheat flour. 2

Pass this flux through a 40-mesh sieve, and mix it as directed under 2.

For one assay ton ore, take 90 grams of this flux, and cover the charge with 15 grams unfused borax. For one-half assay ton ore, take 50 grams of this flux, and cover the charge with 10 grams unfused borax.

CHAPTER II.

3. To Prepare the Sample of Ore.

Take several pounds of rich silicious (quartz) ore, free from sulphur, arsenic, antimony, etc., break the large pieces by a hammer, then in an iron mortar (Fig. 4), so

FIG. 4.

that the largest pieces are not over about an inch through the longest diameter, then pass the ore through the crusher (Fig. 5), then shovel the ore on the tin sampler (Fig. 6),

5

where half is caught and half passes through,

Fig. 5.

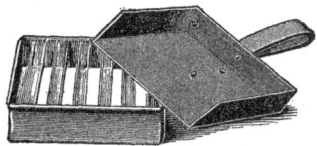

Fig. 6a.   Fig. 6.

repeating until the sample is reduced to about a pound. Now pass it through the

crushing rolls, and pulverize it on a bucking-board (Fig. 7) until it passes an 80-mesh sieve (some use a 100-mesh).

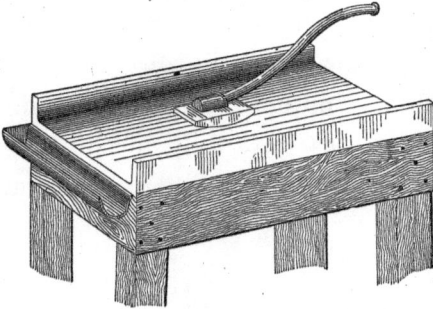

FIG. 7.

In conducting the operation on the bucking-board, place your left hand on the muller and grasp the handle with the right hand; throw your weight on the muller and move it back and forth, raising the handle when drawing toward you, and depressing it when pushing backward.*

---

*Ores can be pulverized in the iron mortar alone, or by other apparatus than is mentioned here.

Everything used in these operations must be perfectly clean. Particles of rich ore left from the last sample in the iron mortar, in the crusher, on the bucking-board, in the

After passing the ore through an 80-mesh
sieve, mix it thoroughly on a rubber or an
oil cloth. Spread the ore out evenly on the
cloth, lift a corner of the cloth and draw it
forward, rolling the ore on itself (merely
sliding the ore on the cloth does not mix it),
and proceed as directed for mixing flux
(under 2), and put the sample into an ore-

FIG. 8.

pan (Fig. 8). Read the notes on ore-sam-
pling and ore samples on page 29.

4. Assay Ton Weights.

The ore is weighed by Avoirdupois, and
the gold and silver by Troy weight. The

---

sieve, in the cleaning-brush, or on the mixing-cloth, care-
lessness or uncleanliness in any of the operations or about
the laboratory, make the results unreliable; and unless a
person is strictly honest and is determined to do accurate
work, he should not engage in this line of work.

After pulverizing rich ore, clean the apparatus with
barren quartz or glass.

system of assay ton weights is very convenient and easily understood.

1 lb. Avoirdupois = 7000 Troy grains.

2000 lbs. = 1 ton.

2000 × 7000 = 14,000,000 Troy grains in 1 ton Avoirdupois.

480 Troy grains = 1 oz. Troy.

14,000,000 ÷ 480 = 29,166 + Troy ozs. in 2000 lbs. Avoirdupois.

By taking as many milligrams of ore as a ton contains ounces, every milligram of gold or silver extracted is equivalent to an ounce to the ton. In one assay ton (A. T.) there are 29,166+ milligrams. Hence by taking one assay ton of ore, 1 milligram of gold or silver extracted = 1 oz. Troy to the ton of ore. 2000 lbs. : 1 A. T. :: 1 oz. Troy : 1 milligram.

5. To Make the Assay.

Weigh an assay ton of the sample, prepared as directed under 3, on one of the pulp-

balances * (Figs. 9 and 10). Mix the sample
of ore thoroughly, spread it out in a thin
layer, then take the sample, with a steel
spatula having a straight end (Fig. 11), from
all parts of the layer, driving the spatula to
the bottom of the layer each time. This

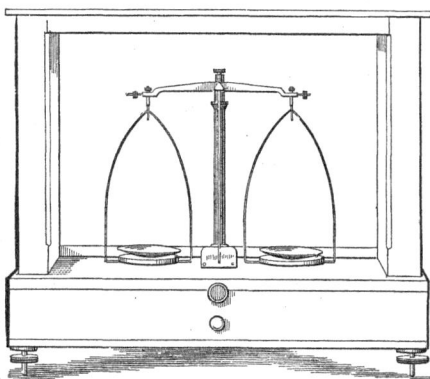

FIG. 9.

should be done gently, taking care not to jar
the ore. Take about 85 grams, or 64 cc.,
of the flux (Chap. I, 2)†, and mix the ore
and flux in the same manner as directed for

---

* Put the assay ton weight on the left-hand pan.
† Or use the flux given on page 4, as there directed.

mixing sample (under 3), then pour it into a
20-grm. clay crucible (Fig. 12), in which

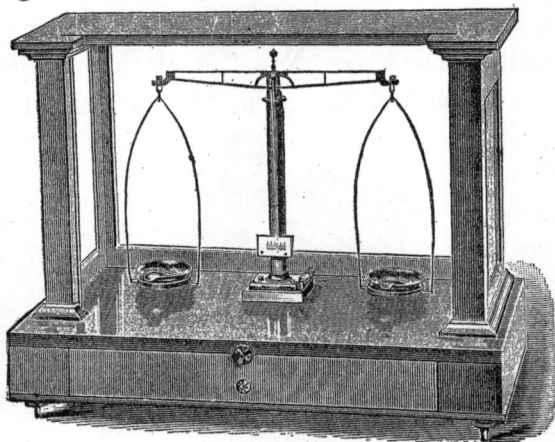

FIG. 10.

FIG. 11.

FIG. 12.

about 10 milligrams of C. P. silver has previously been placed. Weigh out another assay
ton of the ore and proceed as before, omitting

the silver.* Cover the charge in each crucible
with about 15 grams of unfused powdered
borax. Consider the charge in the crucible
containing the silver No. 1, and the other
No. 2. Place No. 1 on the left in the cru-

FIG. 13.—Gasoline Crucible-furnace.

cible-furnace † (Fig. 13), and No. 2 on the
right. You will know them by these positions through all the operations.‡

---

* When assaying for gold only, add silver to both charges.
When the approximate amount of gold in the ore is known,
add an amount of silver to insure separation (see 8).

‡ Crucibles and scorifiers can be marked with reddle.

† If a gas-furnace is used, the crucibles may be put into
the furnace before the fire is started. If a coke- or a coal-
furnace is used, the fuel should be red hot and in such a
condition, before the crucibles are put into the furnace, that
the heat can be raised gradually. Make a nest in the glow-
ing coals for the crucible, and pack the fuel around the crucible. Or put an old crucible into the fuel, and put the cru-

6. Fusion in the Crucible.

Fuse the charge, keeping the fire hot enough to keep the charge liquid. When the charge has subsided and is in a state of quiet fusion (the chemical reactions having all taken place, the charge may still have a con-

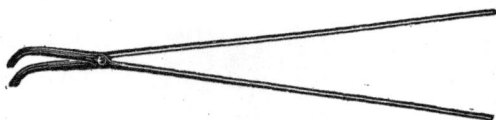

vection motion), urge the fire and after a few minutes remove the crucibles, one after another, from the fire, with the crucible-tongs (Fig. 14), give the crucible a circular, swinging movement, to wash the sides of the crucible, tap it on the floor or table to settle the

cible with the charge in its place, when the fire is in the proper condition. If, after the charge melts, the heat is checked sufficiently to stop the action, the assay is said to "freeze." In this case the assay is defective, and a new assay must be made.

Crucibles with the charges should be gently heated, before they are put into a hot fire, to prevent the crucibles from breaking, and to prevent "blowing" (see under sodium bicarbonate on page 48).

·lead, and pour the liquid contents, at first slowly, then rapidly, into a mould* (Fig. 15), which has previously been warmed.

FIG. 15.

When the assay has partly cooled, turn it out of the mould.   The slag should be evenly colored, and should contain no glob. ules of lead.   Break the slag by a

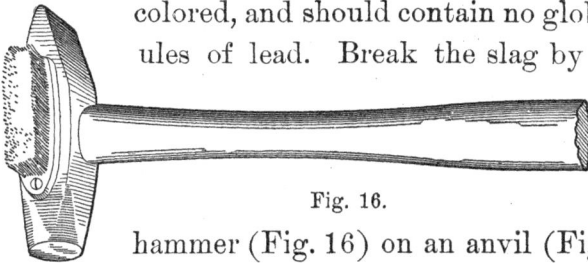

Fig. 16.

hammer (Fig. 16) on an anvil (Fig. 17), hammer the lead button to a cube, flatten the corners, brush it (Fig. 18), and place it in its proper place on the button-tray (Fig. 19).

---

* Some assayers pour off carefully the slag into a mould until the lead button is exposed, then they pour the remaining slag and the lead into a separate mould. By this method the button cools more quickly; and, if the lead contains sulphur (looks coarse and fuming), the crucible can be returned to the furnace, and the sulphur allowed to burn out, before pouring the lead.

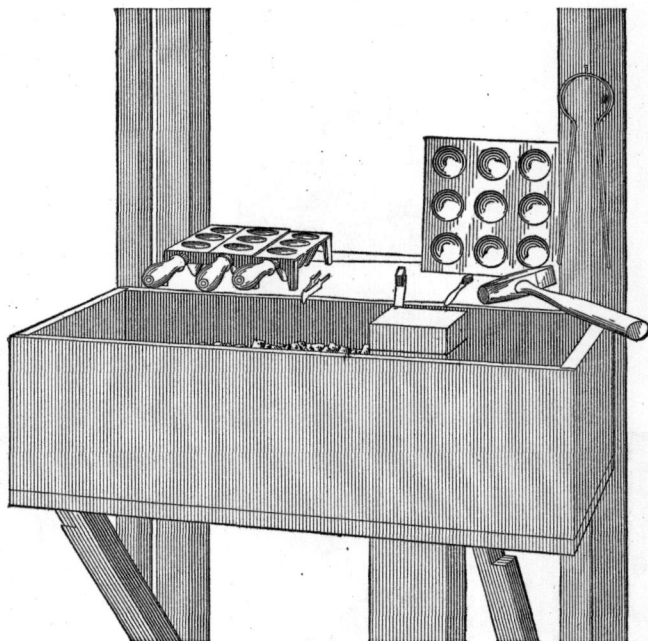

Fig. 17.

Fig. 18.

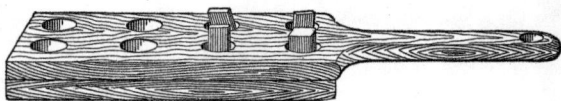

Fig. 19.

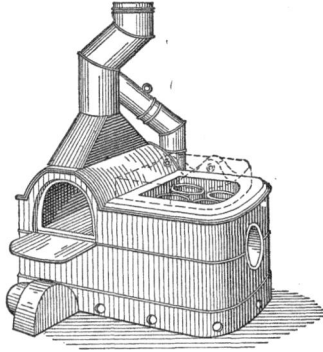

GASOLINE COMBINED CRUCIBLE AND MUFFLE-FURNACE.

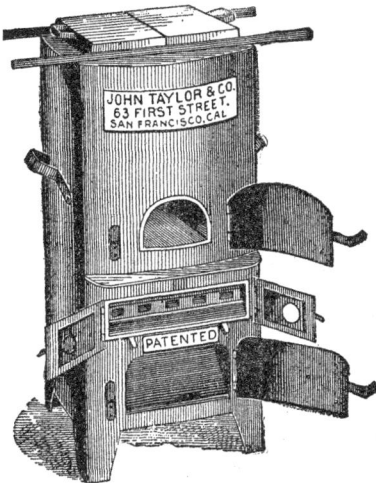

FIG. 20.

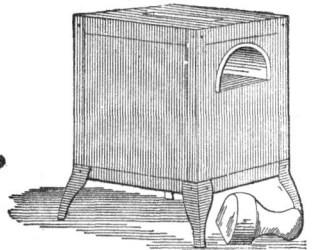

COKE-BURNING COMBINED CRUCIBLE-
AND MUFFLE-FURNACE.

GASOLINE MUFFLE-
FURNACE,

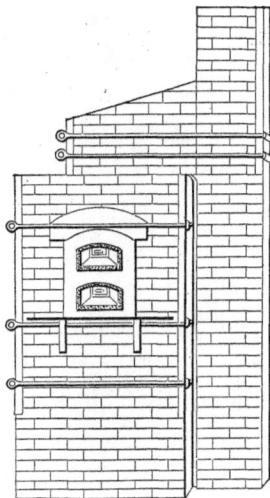

SOFT-COAL-BURNING MUFFLE-FURNACE.

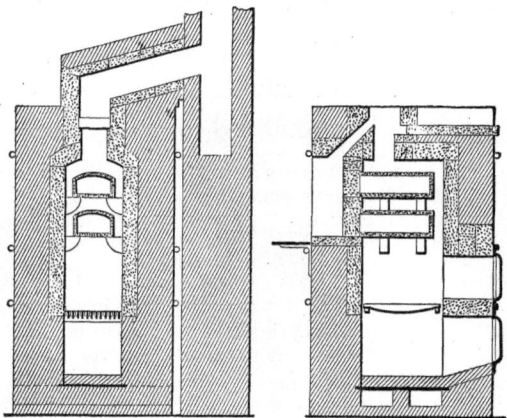

TWO SECTIONS OF THE ABOVE.

7. Cupellation.

Place cupels in the muffle of the muffle-furnace (Fig. 20) in the same relative positions that the crucibles occupied in the crucible-furnace.   Place an extra cupel in front of each row of cupels.   A little bone-ash should be sprinkled on the floor of the muffle.

When the cupels are at a red heat, and the muffle at a bright orange-red,* charge in the buttons by the cupel-tongs (Fig. 21), and close the muffle until the black crust has disappeared from the melted buttons, then open the muffle.   Now lower the temperature † sufficiently to form litharge crystals, "feathers," on the cupel.   The lead oxidizes, and, together with the oxides of other metals, is absorbed by the cupel.   When nearly all the lead has been oxidized, moving rainbow-

---

* An idea of the approximate temperature in the muffle can be gained by Pouillet's scale of temperatures:

	Deg. C.	Deg. F.		Deg. C.	Deg. F.
Incipient redness.	525	977	Deep orange....	1,100	2,012
Dull red........	700	1,292	White..........	1,300	2,372
Cherry-red	900	1,652	Dazzling white.	1,500	2,732

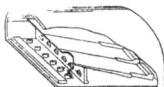

LONGWOOD'S CUPEL
COOLER.

† The temperature can be lowered by checking the fire, or by putting cold crucibles, scorifiers, or other muffle coolers into the part of the muffle that is too hot. replacing them, when hot, with cold ones until the desired temperature is obtained.

colored rings appear on the button, **after** which the button becomes duller, and **the**

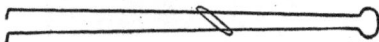

FIG. 21.

cupellation is finished.* If the button is less than one-third gold, it should be cooled slowly by placing muffle-coolers near it in the same manner as directed for cooling the muffle on page 18, or by placing a red-hot cupel over it, as it absorbs oxygen when molten and gives it out suddenly on solidifying, from which loss may result from "spitting."

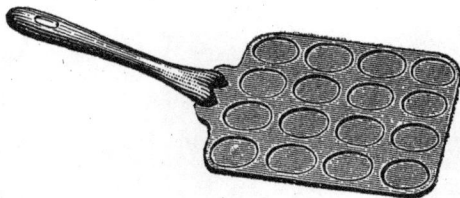

FIG. 22.

Remove the cupels from the muffle, placing them on a tray (Fig. 22) in their proper

---

* Near the end of the cupellation, before the colored rings appear, raise the temperature.

positions.　When the buttons have partially
cooled, remove No. 2 from the cupel by means

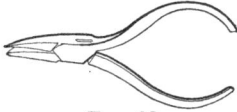

FIG. 23.　　　　　　　　　FIG. 24.

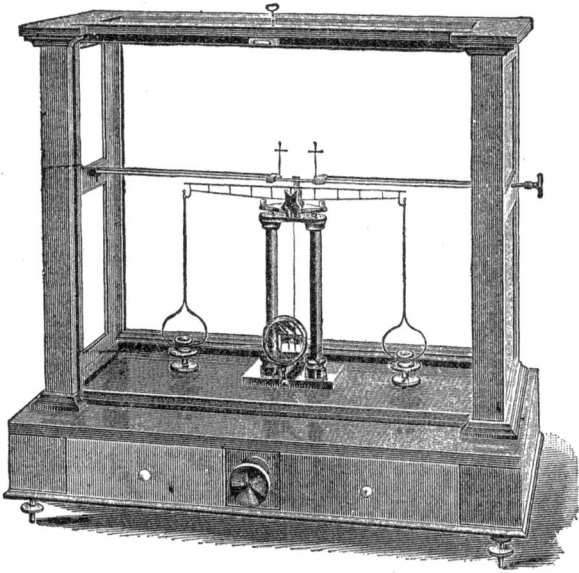

FIG. 25.

of a pair of pliers (Fig. 23), squeeze it to
loosen the adhering bone-ash, and brush it

with the assay-button brush (Fig. 24), place
it on the left-hand pan of the button-balance
(Fig. 25), and weigh it.

8. Parting.

After noting the weight of the button, fuse
it with sufficient silver to make the silver
about $2\frac{1}{2}$ times that of the gold,* lay it on an
anvil and flatten it by a few blows of a hammer. Pour nitric acid, 1.16 sp. gr. (21°
Baumé) into a test-tube (Fig. 26) to the
depth of a little over the diameter of the test-
tube, warm the acid, and then put the bead
into it, and boil it until the acid becomes
colorless. After all fine particles have settled,
pour off the acid and boil again with the
same amount of acid of 1.26 sp. gr. (32°
Baumé). Pour off the acid, fill up the test-
tube with distilled water, and after the gold
has settled, pour off the water, fill the

---

* This is called "inquartation."

test-tube again with water, invert an annealing-cup (Fig. 27) over the mouth of the test-tube, and, by a quick movement, invert the test-tube, keeping the cup over the mouth of the test-tube, and fill the cup with water. The gold settles into the cup. Raise the

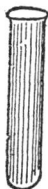

test-tube a little at a time until it is nearly even with the top of the annealing-cup (this should be done over a sink), then (after the gold has settled) quickly remove the test-tube, and after the water has run out, hold it to the top of the cup (using it as a pouring-rod) and pour the water out of the cup. With the last drop, by tapping the cup, bring all the gold together in the bottom of the

cup. Heat the cup to redness, and put it
on the tray (Fig. 28) in its proper place.
Do the same with the button, No. 1, (except that this button may need no additional
silver). When cool, weigh No. 2. The difference between the weight of the button and
the weight of the gold is the weight of the

FIG. 28.

silver. Weigh the gold of No. 1, which
should closely agree with the weight of No. 2.

Read the notes on Fusion in the Crucible
and on Cupellation on pages 32 and 34, and
repeat the assay until you can make duplicate assays (duplicates by adding the same
amount of silver, and duplicates without
adding silver) that agree. Then practise on
low grade ores of similar character.

9. Scorification Assay.

From the sample of ore prepared for the crucible assay (under 3), take $\frac{1}{10}$ assay ton (A. T.), weigh out 1 A. T. of C. P. granulated lead, mix about half of the lead with the ore, put the mixture into a scorifier (Fig. 29), smooth it down, mix the remaining lead

Fig. 29.

on the cloth or paper on which the ore and lead were mixed, so as to take up any particles of ore that may have been left on the mixing-cloth, spread it over the mixture evenly, and put about 0.1 gram of borax glass on top. Add a little silver to this, and make a duplicate, omitting the silver.*

The scorification assay is made in the muffle, which should have a temperature of 1050° to 1100° C. When the muffle has reached

---

* When assaying for gold only, add silver to both charges.

the required temperature, charge in the scorifiers * by the scorifier-tongs (Fig. 30), in the
relative positions as directed for the crucibles
under 5. Close the muffle until fusion takes
place, then open it. A ring of slag soon

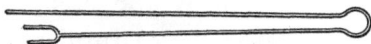

FIG. 30.

forms which is increased by the addition of
the lead oxide which forms, and in about 30
or 40 minutes, it closes over the metals.
When the slag has closed over the metals,
place, by the cupel-tongs, about 0.2 gram of
powdered charcoal, wrapped in tissue-paper,
on the surface of the slag, and close the
muffle. This reduces some of the lead oxide,
and the globules of lead fall through the
slag and carry the gold and silver down,
which had remained in the slag. When the

* Scorifiers should be dry, and should be warmed with
the charge by placing them on the furnace for some time
before charging them into the muffle, to prevent them from
breaking.

fusion becomes quiet, pour it into a mould. The lead buttons are treated in the same manner as already explained under the crucible assay.

Read the notes on the Scorification Assay on page 43, and repeat the assay until you can make duplicate assays (duplicates by adding the same amount of silver, and duplicates without adding silver) that agree.

10. Lead Flux.

	Per cent. by Weight.
Sodium bicarbonate	50
Potassium carbonate	25
Borax-glass	12
Flour	9
Silica (powdered sand)	4

Pass through a 40-mesh sieve, and mix as directed for mixing flux, page 3.

11. Fire Assay for Lead.

Prepare a sample of galena (lead sulphide) ore as directed under 3. Weigh out 5 grams of the ore * and mix it with 20 grams of lead flux, pour it into a 5-gram clay crucible (Fig. 31), cover it with dry fine common salt † to the depth of about ¼ inch, and stick 3 8-penny nails into the charge. Make a duplicate assay.

First Method.—Make this assay in the muffle,‡ which should be at a cherry-red heat before the crucibles § are charged in. After the crucibles are introduced into the muffle,

---

* For important work take 10 grams ore, 35 grams flux, a 10-gram crucible, a cover of common salt (about 30 grams), and stick 3 10-penny nails into the charge. Take 20 grams of tailing or low-grade ore for an assay.

† Mix the salt on the cloth on which the ore and flux were mixed, so as to take up any particles of ore that may have remained on the cloth.

‡ The opening in the back part of the muffle should be closed with fire-clay while assaying for lead.

§ See last paragraph of footnote on page 13.

close the muffle, and, after the effervescence, "boiling," in the crucibles has commenced, lower the temperature sufficiently to allow the effervescence to go on slowly, and, after the effervescence ceases, which will be in 30 to 40 minutes, if the temperature is properly regulated, raise the temperature. In 45 to 50 minutes from the time that the crucibles

FIG. 31.   FIG. 31*a*.

were introduced into the muffle the fusion will be complete; then take out one crucible after another and remove the nails with a pair of small tongs (Fig. 31*a*). Seize the nails above the slag and throw the lead off the nails by tapping the tongs on top of the crucible, give the crucible a circular, swinging movement, to wash the sides of the crucible,

·tap it on the table to settle the lead, and then pour the assay into a mould.   After cooling, hammer the slag off, flatten out the lead button, and weigh it.

Second Method.—This method is the same as the first method, except that the assay is run at a lower temperature and for a longer time. The muffle should be at a low-red heat when the crucibles are charged in.   The temperature should not be allowed to go above redness. If the blue flame above the crucible "jumps," the temperature is too high, and it should be lowered immediately.   After the crucibles have been in the muffle an hour and a quarter to two hours, if the blue flame has disappeared, raise the temperature.   After some time (the length of time depends upon the regulation of the temperature) the assay will be in a state of quiet fusion, then take out the crucibles and proceed as directed above in the first method.

Example.—If the lead button weighs 3 grams, $3.00 \div 5$ (the amount of ore taken) = 60 per cent of lead ; 60 per cent of 2000 lbs. = 1200 lbs. lead in a ton of ore. When lead is worth 4 cents a pound, 1200 lbs. are worth $1200 \times \$0.04 = \$48.00$.

CHAPTER III.

12. Notes on Ore-sampling and Ore Samples.

1. Large samples of ore should be crushed to a size not much larger than ¼ inch, mixed on a cloth, spread out evenly, and divided into quarters. Remove two opposite quarters, mix the other two thoroughly, spread out, and quarter as before. Repeat this until the sample is reduced to the required size. After the sample is reduced to 2 or 3 lbs., it should be crushed finer before reducing it to a smaller sample.

Sampling can also be done by *channelling*, which consists in spreading the mixed ore out in a square, taking out samples in parallel grooves across the square, as far apart as the

29

width of the channels, then at right angles to these channels; or by a sampling tin, quartering-shovel, split shovel, or mechanical sampler.

2. Before pulverizing, damp samples must be dried on a water-bath (Fig. 32), or in a drying oven (Fig. 33) at a temperature not above 100° C.

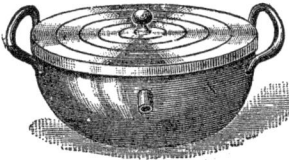
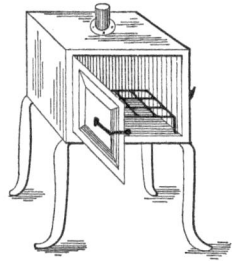

FIG. 32.   FIG. 33.

3. After the sample is thoroughly mixed, pour it into an ore-pan.   Mark the number of the sample, the character of the ore, and other data on a piece of paper or cardboard and put it into the pan with the ore.

4. Do not shake, tap, or otherwise disturb

the sample before weighing out the amount
of ore for assay.   If the ore is not immediately weighed out, the sample must be mixed
again before weighing out the ore for assay.
By standing, the heavy particles have a tendency to settle to the bottom of the pan on
account of their weight, aided by the vibra-
tions of the building, produced by various
causes.

5. All the ore should pass the sieve. If
scales of gold are left on the sieve, put them
on the bucking-board, cover them with part
of the fine ore and grind hard. Repeat until
they pass the sieve. If the scales cannot be
pulverized, weigh the sample that passes the
sieve, and the scales separately, cupel the
scales with a little lead-foil, and calculate the
value per ton of ore, to which add the
value of the assay of the ore that passed the
sieve.

For methods of assaying ores and materials
containing scales of gold, or other coarse

metal, and for a method for calculating the value of such ores, see Chapter VIII.

13. Notes on Fusion in the Crucible.

1. The ore and flux must be intimately mixed, so that as the lead is reduced it can come in contact with the gold freed by the pulverization, and by the fusion of the ore.

2. The object of the fusion is to collect the gold and silver in a button of lead, reduced from the litharge, and to form a fusible slag with the fluxes and gangue of the ore.

3. Crucibles should not be over three-fourths full, and less in case of sulphide ores.

4. The assay swells when heated; prevent it from boiling over.

5. Causes of boiling over: (*a*), too much borax, especially in oxidized ore; (*b*), too much soda; (*c*), unfused borax mixed with the charge; (*d*), rapidly heating a charge containing nitre.

6. Boiling over can be prevented by re-

moving the cover, if covered, checking the heat, or by throwing a teaspoonful of salt into the crucible.

7. If the slag is pasty, add borax. If accuracy is required, make another assay, fluxed as indicated by the first assay. The fluxing of other than silicious ores will be explained when they are taken up.

8. After pouring, examine the crucible. If the crucible contains shots of lead or pasty masses, the assay is defective.

9. The fusion can be made in a muffle; the temperature should be about the same as in the scorification assay. When the fusion is made in the muffle, $\frac{1}{2}$ an assay ton of ore is usually taken, and fluxes in the same proportion.

14. Notes on Cupellation.

1. The cupel should be about twice as heavy as the button.

2. Spirting of the lead button sometimes happens when there is too strong a draught, or when the button contains volatile elements, as arsenic, antimony, sulphur, carbon, etc. If small beads of metal are found on the cupel, the assay is defective.

3. When the muffle is not hot enough, put a piece of charcoal in the front of the muffle, which will raise the temperature. Some also use this to equalize the temperature in the muffle, as the front of the muffle has always a lower temperature. The temperature increases from the front to the back of the muffle.

4. After the cupellation is finished, the button should be left in the muffle a few minutes to remove all remaining traces of lead.

5. When the button begins to solidify, it "flashes," suddenly "brightens," which is due to the sudden disengagement of the latent heat of fusion.

6. Buttons containing a large proportion of silver will "spit," if cooled rapidly. This "spitting," "sprouting," or "vegetation," as it is called, may be prevented in three ways: (*a*) by placing cold crucibles, scorifiers, or other cupel coolers near the button and replacing them, when hot, with cold ones until the button has solidified. A cupel should not be moved while the button is liquid, unless it can be done without moving the button from its place on the cupel, or loss will result; (*b*) by inverting a red-hot cupel over the cupel and button; (*c*) by closing the muffle, withdrawing the fire, and allowing slow cooling to go on. The "spitting" is due to the absorbing of oxygen by the silver when molten, and giving it out suddenly when solidifying.

7. Silver is volatile at a high temperature, also gold, to some extent. Silver begins to volatilize at a white heat.

A strong draught of air cools the cupel, and prevents the oxides from being absorbed as fast as formed.

The longer the time required for cupellation, the more the loss by volatilization and absorption by the cupel.

When the buttons are charged into the muffle, the temperature should be high enough to start a rapid oxidation, then the temperature should be lowered sufficiently to form litharge crystals on the cupel. Near the end of the cupellation, the temperature should be raised or the lead may not be completely removed from the buttons.

8. Sometimes during cupellation the action stops, and the button solidifies. This is called "freezing," and is due to the oxides forming more rapidly than they can be absorbed, or to the low temperature of the

muffle. If the former, an addition of lead will remedy it. If the latter, raise the temperature. The results are not reliable.

15. Notes on Parting.

1. When the bead contains from two to three times as much silver as gold by weight, nitric acid will dissolve the silver. This separation is called "parting." Not all the silver will dissolve, but it is sufficiently accurate for ordinary work.

2. If the button is silver white, it contains more than 60 per cent silver. If it is yellow, it needs an addition of from 2 to 2½ times its weight of silver (according to the degree of yellowness), to part it. The button and the silver may be fused by the blowpipe on charcoal, or by wrapping them in lead-foil and cupelling. Before parting the button, it is flattened by a hammer, and if large, passed through the rolls (Fig. 34),

after which it is annealed by heating it to redness.

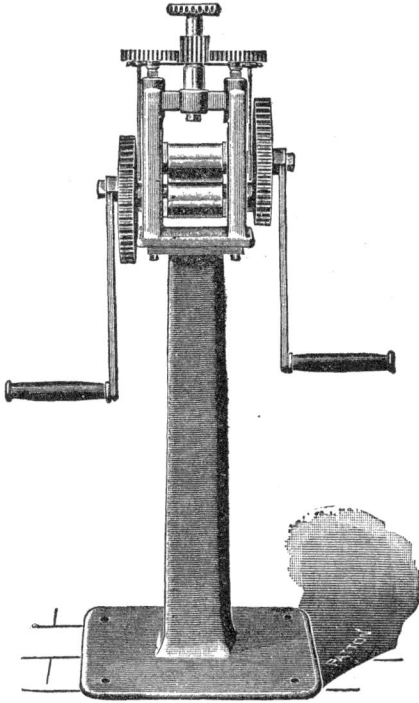

FIG. 34.

Silver fuses at 954° C., and gold at 1045° C. The silver melts first and the gold sinks in it.

When the fusion is made by the blowpipe, care must be taken to continue the heat until the gold is fused.

3. Parting can be effected in a test-tube, in a small porcelain crucible, or, if the button is large, in a parting-flask (Fig. 35). If the separation is made in a porcelain crucible, the

FIG. 35.

gold is also dried and annealed in the same crucible. In annealing, the crucible should be heated to redness. The heat must be raised gradually to prevent the gold from being scattered by the steam formed.

When the separation is made in a porcelain crucible, after pouring off the wash-water, bring the gold together, and turn the crucible so as to collect the remaining water on the

opposite side from the gold.   The water can then be removed by filter-paper.

A needle, stuck into a soft piece of wood for a handle, is convenient to loosen the gold from the cup, after annealing.   The gold can be weighed in this form, or some wrap it in lead-foil and cupel it.

4. Use dilute acid, 1.16 sp. gr. (21° Baumé), for the first boiling, and fresh acid of 1.26 sp. gr. (32° Baumé) for the second boiling.   If the acid is warmed to 90° before the bead is put into it, the gold does not break up into such fine particles.   By boiling a second time, the silver is more completely removed, the gold becomes more compact, and there is less danger of loss.   By boiling the acid after it is concentrated to about 1.42 sp. gr., it dissolves gold in appreciable quantities.

To prevent violent action and so keep the gold from breaking up into fine particles, after heating the dilute acid to nearly boiling, drop the bead into it and continue to heat it

on a sand-bath or hot plate for ten or fifteen minutes; pour off the acid, add stronger acid, boil and proceed as directed above.

5. If the gold, after parting, weighs more than one-third of the weight of the bead, the gold contains silver and must be fused again

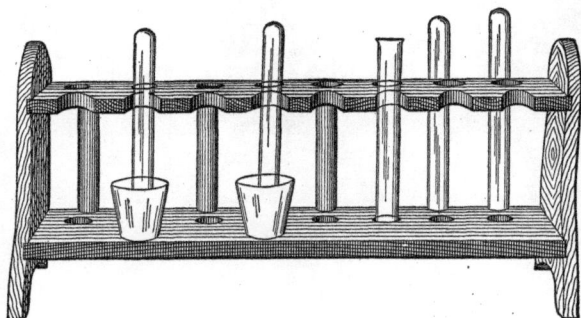

FIG. 36.

with from 2 to 3 times its weight of silver, and again parted.

6. If from $2\frac{1}{4}$ to $2\frac{1}{2}$ times as much silver as gold is present, the gold remains in one piece after parting; if much more silver is present, the gold breaks up into small particles, some of which may float on the surface of the acid.

By touching them with a glass rod, or by dropping water on them, they can be made to sink. Set the test-tube in the rack (Fig. 36) until all the gold has settled, and bring the particles together by tapping the bottom of

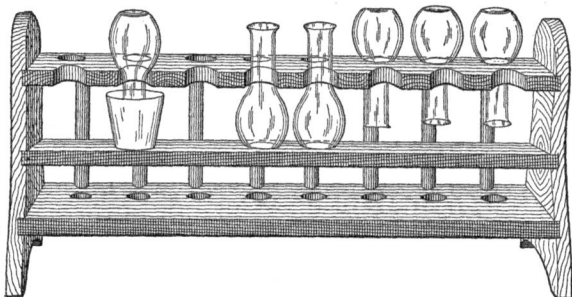

FIG. 36*a*.

the test-tube with the fingers, before pouring off the acid.

7. If, in beginning to boil, the bead turns black and the action stops, it must be fused with additional silver.

8. Small glass tubing in the acid while boiling will prevent "bumping."

16. Notes on the Scorification Assay.

1. This method is especially applicable to rich gold and silver ores. The scorification assay gives higher (uncorrected) results for silver.

2. If the slag is pasty, add borax. The slag may be rich.

3. Borax makes the slag fluid, but if too much borax is used, the slag will cover the bath of metal too soon.

4. If the muffle is not hot enough, when the scorifiers are charged in, gold may remain in the slag.

5. Pasty slag or slag from rich ores should be ground up, fluxed with a little borax and flour, or argol, and scorified as the ore. Use the same scorifier in which the first scorification was made, and cupel the button with the first one obtained.

6. If the ore is low grade, make a number of assays, place the buttons together in a

scorifier, add a little borax glass, and scorify them to the proper size for cupellation. In this case the slag need not cover the bath of metal.

7. The treatment of other than silicious ores, and ores containing other elements, will be explained when they are taken up.

17. Notes on the Fire Assay for Lead.

1. The fire assay for lead gives only approximate results. If the ore contains gold, silver, iron, copper, or other metals, some or all may be reduced with the lead. If the button is brittle, it contains antimony, sulphur, etc.; if hard, copper, iron, etc. Arsenic carries lead into the slag, and forms with the iron a separate, hard, and brittle button. Some lead may volatilize, and some oxidize and go into the slag. The button can be cupelled for the precious metals and the weight deducted.

2. If the assay is made in a crucible-furnace, the crucible should be covered.

Avoid too high a temperature. Lead and lead sulphide are volatile at a high temperature. The heat should not be above redness, at least for the first 30 minutes.

3. Ores free from sulphur need no nails.

4. The assay can be poured, or allowed to cool in the crucible, after which the crucible is broken and the lead extracted.

5. The slag can be assayed, and the lead recovered, added to the first button.

6. Lead fluxes :

Mix and take from 18 to 20 grams for 5 grams of ore, and to ores containing sulphur, add 3 8-penny iron nails, or use a wrought-iron crucible. With ores containing phosphorus, use additional borax glass to prevent pasty slag.

		(1)	or	(2)	or	(3)	
1.	Sodium bicarbonate......	16		16		100	parts
	Potassium carbonate.....	16		12		50	"
	Wheat flour..	4		12		15	"
	Borax glass........	8		4		25	"
	Salt....................cover			cover		cover	

		(1)	or	(2)	or	(3)	
2.	(Plattner.)						
	Potassium carbonate.........	5		$6\frac{1}{2}$		2	parts
	Sodium bicarbonate..........	$6\frac{1}{2}$		5		2	"
	Flour.....................	$2\frac{1}{2}$		1		1	"
	Borax glass	$2\frac{1}{2}$		$2\frac{1}{2}$		1	"
	Salt .... cover	cover	cover				

3. Ore .... 10 grams
   Potassium cyanide................ 35 "
   Salt............................. cover
   No nails.

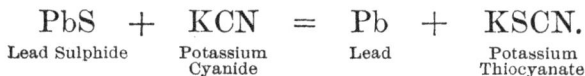

$$PbS + KCN = Pb + KSCN.$$

Lead Sulphide   Potassium   Lead   Potassium
                Cyanide   Thiocyanate

Potassium cyanide is a strong reducing agent, but, if used with ores containing iron, copper, and other metals, it reduces them also with the lead.

Potassium cyanide is extremely poisonous.

Combination Method.—Make the assay in the usual way, using a suitable flux (not KCN) for the ore, but only one nail. About

5 minutes after complete fusion of the charge add about 10 grams potassium cyanide. If the cyanide is fine, remove the crucible quickly from the muffle, add a rounded teaspoonful of the cyanide, replace the crucible, and in about 15 minutes the charge will be ready to pour. After adding the cyanide, avoid a high temperature or the charge will boil over.

This method was worked out by Mr. Oscar J. Frost.

CHAPTER IV.

Sodium Bicarbonate, $NaHCO_3$, or its corresponding potassium salt, $KHCO_3$, acts as a basic flux, a desulphurizing agent, and in some cases, as an oxidizing agent. Heat reduces it to sodium carbonate, which on being fused with silica, forms fusible silicates, and liberates carbon dioxide. The carbon dioxide liberated oxidizes sulphur, and metallic iron, zinc, and other metals, which then go into the slag. With sulphide ores, under certain conditions, it forms sodium sulphate and sodium sulphide.

$$2NaHCO_3 + \text{Heat} = Na_2CO_3 + CO_2 + H_2O.$$
166.86 *

---

* The atomic weights from the sixth annual report of the committee on atomic weights are used in this book.

A charge containing sodium bicarbonate should be heated gently at first, or the carbon dioxide and steam, which are given off at a comparatively low temperature, may blow out some of the fine ore. The carbonate fuses at about 800° C., and absorbs such infusible substances as lime, alumina, etc.

When equal molecules of silica and sodium carbonate are fused together, sodium bisilicate is formed, and carbon dioxide liberated.

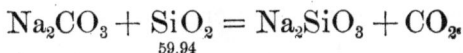

$$Na_2CO_3 + \underset{59.94}{SiO_2} = Na_2SiO_3 + CO_2.$$

When fusion takes place, the silica displaces the CO_2 in the sodium carbonate. The escape of the CO_2 causes a brisk effervescence to take place.

According to the above equations, it takes over 2.78 times as much sodium bicarbonate as silica to form sodium bisilicate.

Sodium carbonate forms a double salt with potassium carbonate, which is more fusible and fuses at a lower temperature than either alone.

When silica, sodium carbonate, carbon, borax, and other substances are fused together, as in an assay, of course many other reactions take place.

Litharge, PbO, is used in the crucible assay of gold and silver ores. It is a basic flux, dissolves metallic oxides, oxidizes sulphur and other oxidizable substances, and supplies the lead in the crucible assay.

Nearly all litharge contains silver, the amount of which must be determined, when used in the assay for silver. It can be freed from red lead, Pb_3O_4, by fusing in a crucible, pouring, and keeping it from the air while cooling. It is claimed that red lead oxidizes silver and thus causes loss.

With silica it forms a lead silicate.

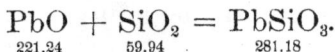

$$PbO + SiO_2 = PbSiO_3.$$
$$\underset{221.24}{} \quad \underset{59.94}{} \qquad \underset{281.18}{}$$

It takes over 3.69 parts of lead oxide to one part of silica by weight to form lead bisilicate, which is easily fusible, and more fluid, when fused, than sodium bisilicate. It forms fusible double silicates, but has no action on lime or magnesia except in the presence of silicates and borates.

Metallic oxides that are infusible or almost infusible alone are easily dissolved by lead oxide, with which they form a basic slag, which attacks the crucible by uniting with the silica in the crucible.

Borax, $Na_2B_4O_7$.—The unfused or crystallized borax, $Na_2B_4O_7$, $10H_2O$, contains over 47 per cent water. When mixed with the charge of ore and flux, the fused borax (borax glass) should be used, as the unfused borax is apt to cause loss by swelling and

thus forcing part of the charge out of the crucible. It is an acid flux, and will flux sulphides, arsenides, metallic oxides, lime, etc. It also forms fusible compounds with silica.

Silica, SiO_2, is an acid flux. It forms compounds with all the bases, and is used in the assay of ores containing lime, magnesia, baryta, iron, etc. Pulverized silica (pure white quartz) should be used.

As a substitute, ground glass can be used. Window-glass is probably the best, which is a silicate of sodium and calcium. Bohemian glass is a silicate of potassium and calcium, and is not easily fusible. Bottle-glass is a silicate of sodium, aluminum, calcium, iron, etc. Flint glass contains lead, and hence cannot be used in the assay for lead. Enamelled glass should not be used, as it contains arsenic, antimony, tin, etc.

Flour.—Wheat flour is a strong reducing

agent. One part of flour will reduce about 15 parts of lead from lead oxide.

Argol is a reducing agent, and basic flux. It contains other carbonaceous matters than the pure potassium bitartrate, $KHC_4H_4O_6$, and therefore has a greater reducing power. Heat changes it to potassium carbonate and carbon. Its reducing power can be determined by fusing it with litharge and sodium bicarbonate. One part of argol will reduce from 5 to $8\frac{1}{2}$ parts of lead.

Potassium Cyanide, KCN, is a strong reducing and desulphurizing agent. It combines with oxygen, forming cyanate,

$$PbO + KCN = Pb + KCNO;$$

and with sulphur, forming thiocyanate,

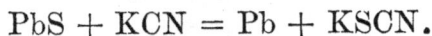

$$PbS + KCN = Pb + KSCN.$$

Potassium cyanide is extremely poisonous.

Charcoal.—Powdered wood-charcoal is used. It is a reducing agent. It absorbs water and gases, and has about 3 per cent ash, $2\frac{1}{2}$ per cent hydrogen, $1\frac{1}{2}$ per cent oxygen, and 93 per cent carbon. At low temperatures it forms carbon dioxide; and at high temperatures, carbon monoxide, when heated with metallic oxides, from which it takes the oxygen. As its reducing power varies under different conditions, its reducing power should be determined under the conditions under which it is to be used. Ordinarily 1 part of charcoal reduces from 20 to 30 parts of lead.

Iron in the form of nails and wire is used in the assay of sulphide ores. It forms iron sulphide with the sulphur, which is dissolved by the slag, if only a moderate quantity is present. A large quantity will form a separate layer of matte.

Potassium Nitrate, KNO_3 (nitre, or salt-

petre), oxidizes most oxidizable substances. It melts at 339° (Person), 352° (Carnelly); and, at a higher temperature, it is decomposed, yielding a large volume of oxygen. The oxygen combines with the sulphur of sulphide ores, and also oxidizes most of the metals except gold and some of the platinum group. One part of nitre oxidizes about 4 parts of lead.

Salt, sodium chloride, NaCl, is used as a cover. In an open crucible, it melts and volatilizes at a red heat, and forms a liquid cover, which excludes the air and prevents loss by ebullition. It washes the sides of the crucible, and forms fusible compounds with silica, antimony, and arsenic.

Metallic Lead, in the form of sheet lead, is used in cupelling beads and in the bullion assay. Granulated lead (test-lead) is used in the scorification assay. As nearly all lead contains silver, its silver contents should be

determined, and the proper correction made, when used in the assay for silver.

Lead acts as a basic flux, and a solvent or collector for gold and silver.

CHAPTER V.

One of the aims in making up an assay charge is to flux the ore so as to produce, when fused, a liquid slag. If the slag under this condition is not liquid, it may retain some of the precious metals. If the charge is of such a character that it fuses, and becomes liquid at a low temperature, its reducing power is diminished; and, if the slag becomes very liquid in the beginning of the fusion, the reduced lead particles may drop to the bottom of the crucible before they come in contact with all the gold in the charge.

Slags are usually made silicates, which are made up of a basic oxide, as sodium oxide, lead oxide, lime, or baryta, combined with silica, which is acid in character.

55

Silicates are classified as subsilicates, mono-silicates, bisilicates, etc., according to the ratio of the oxygen in the base to the oxygen in the silica.

Name.	Sodium Silicate.	Oxygen Ratio.
Subsilicate	$4Na_2O.SiO_2$	2 : 1
Monosilicate	$2Na_2O.SiO_2$	1 : 1
Bisilicate..........	$Na_2O.SiO_2$	1 : 2
Trisilicate.	$2Na_2O.3SiO_2$	1 : 3
Sesquisilicate	$4Na_2O.3SiO_2$	2 : 3

The fusibility of a slag depends on the character of the bases, and on the percentage of silica it contains. The silicates of lime and alumina are the least fusible. Slags of the composition $CaO.SiO_2$ and $4CaO.3SiO_2$ are fusible, and those of a higher percentage of lime are almost or entirely infusible at the temperature of the assay-furnace.

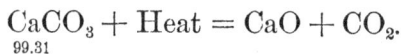

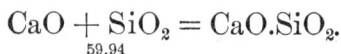

$$CaCO_3 + Heat = CaO + CO_2.$$
$$\underset{99.31}{}$$

$$CaO + SiO_2 = CaO.SiO_2.$$
$$\underset{59.94}{}$$

It takes over 1.6568 times as much limestone, $CaCO_3$, as silica, SiO_2, to produce a slag
of the composition $CaO.SiO_2$; or, to one assay
ton, 29.1666 grams, of limestone, it takes
17.6041 grams of silica.

Neither lime nor silica can be fused at the
temperature of the assay-furnace; but, if
they are pulverized and intimately mixed in
the proportions given above, they unite, at
the temperature attained in the furnace,
forming a fusible compound (calcium silicate) called a *slag*. To this must be added
litharge to supply the lead in which to collect the gold and silver. Sufficient flour or
argol is added to reduce a lead button weighing about 20 grams. Soda, litharge (or potassium carbonate), and silica are added in
such proportions as to make a liquid slag
(when hot), which dilutes the calcium silicate, thus making it more fluid, in order that
the reduced metallic particles of lead, gold,
and silver can sink to the bottom of the

crucible on account of their greater specific gravities.

A charge for a gangue of limestone would be about as follows:

Limestone...............................	½ A. T.
Litharge......................	30 grams
Sodium bicarbonate..........	15 "
Potassium carbonate....................	5 "
Silica......................................	15 "
Wheat flour............................	1 "
Cover of unfused borax..............	10 "

The exact proportions to make a certain slag with each ore cannot be calculated unless a chemical analysis is made of each ore. In assaying, this is out of the question, so the assayer is careful not to add an excess of a flux that is infusible in itself, but he adds an excess of a flux that is very fusible, which, by merely diluting the slag, makes it more liquid.

With magnesia, $4MgO.3SiO_2$ and $2MgO.3SiO_2$ make fusible slags.

$$4MgCO_3 + Heat = 4MgO + 4CO_2.$$
334.60

$$4MgO + 3SiO_2 = 4MgO.3SiO_2.$$
179.82

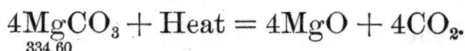

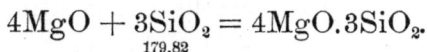

It takes over 1.86 times as much magnesium carbonate as silica to form the slag, $4MgO.3SiO_2$; or, to one assay ton, 29.166 grams, of magnesium carbonate it takes 15.68 grams of silica. For magnesium carbonate the same charge may be used as given above for limestone.

With baryta, $BaO.3SiO_2$ and $BaO.4SiO_2$ make fusible slags.

Silicates of alumina are not fusible alone at the temperature of the assay-furnace. If a base is added, as lime, lead oxide, iron oxide, etc., making them double silicates, they become fusible. A fusible slag can be made by adding sufficient lime and silica to make the

oxygen in the lime between one and two times that in the alumina, and the oxygen in the silica between one-half and twice the sum of the oxygen in the lime and the alumina. Clays are of various compositions, as $2Al_2O_3.3SiO_2$, $Al_2O_3.2SiO_2.2H_2O$, etc.

The most fusible silicates are those of lead, potassium, and sodium, after which come iron and copper. An excess of silica is an advantage in the fusibility of slags in which the base is lime or magnesia, but a disadvantage in those in which the base is sodium, lead, etc. The subsilicates, or basic silicates, are the most fusible; and the fusibility of a slag decreases as the proportion of silica increases, except in lime and other silicates, already noted. Double silicates (silicates having two bases) are usually more fusible than those that have only one base.

CHAPTER VI.

THE ASSAY OF SO-CALLED REFRACTORY ORES.

The assaying of silicious gold and silver ores has been discussed in the previous chapters. When an ore is oxidized, or contains sulphur, antimony, arsenic, tellurium, etc., it needs a different treatment.

Such ores should be fluxed with a view to keep sulphur, arsenic, etc., out of the lead button; and, if one assay ton of ore is taken, to reduce a lead button of from 20 to 30 grams. If the button weighs less than 20 grams, there is danger of leaving some gold in the slag, especially in rich ores. Large buttons give low results for silver, as silver is lost by being absorbed by the cupel, and by volatilization in proportion to the amount of lead present.

The crucible assay is preferred for low grade ores, as a larger amount of ore can be taken. For rich ores, especially for rich silver ores, the scorification assay has advantages.

Ores containing sulphur, antimony, arsenic, etc., have a reducing effect; those containing the higher oxides of iron, copper, manganese, etc., have an oxidizing effect. The former may reduce too large a button; the latter too small a button, or none at all. The reducing action can be corrected by an addition of nitre, and the oxidizing action by an addition of flour, argol, or charcoal. Some ores contain a mixture of oxidizing and reducing agents.

By making a preliminary assay, the proper amount of oxidizing or reducing agents to be added may be calculated. After some experience, the student will know, from the appearance of the ore, about how much of the one or the other reagent to add.

Preliminary Assay.—If the ore contains sulphur, antimony, arsenic, or other reducing elements, take 5 grams of ore, 50 grams of litharge, 18 grams sodium bicarbonate, 5 grams silica, borax cover. Put the charge into a 10-gram clay crucible, and proceed as directed under 6. Detach the lead button and weigh it. Let us assume, for convenience of calculation:

```
1 assay ton................ 30 grams
1 part of nitre oxidizes..... 4 parts of lead (3½ to 4)
Lead button reduced by 5
    grams of ore...... ...... 7 grams
1 assay ton would reduce... 42 grams of lead
This is in excess of the weight
    desired, about........... 20 grams
20 ÷ 4 (1 part of nitre oxi-
    dizes about 4 of lead)..... 5 grams of nitre to be added to
                               one assay ton of ore
```

From this preliminary assay and the chapter on Slags we learn that an assay charge for one-half an assay ton of this ore should be about as follows:

Ore.. $\frac{1}{2}$ A. T.
Litharge...................................... 26   grams
Sodium bicarbonate.................... 15   "
Potassium carbonate................. 10   "
Silica, about.............................. 2.5 "
Nitre, KNO_3............................... 2.5 "
Cover of unfused borax............ 10   "

An assay charge that contains much nitre is liable to boil over. To prevent boiling over: (*a*) start with a low heat and raise the temperature slowly; (*b*) cool the charge by placing cold sheet iron, fire-clay tiles, or other coolers over the crucibles, replacing them, when hot, with cold ones (do not "freeze" the assay. See footnote, p. 13); (*c*) throw a teaspoonful of common salt into the crucible (see also 5 and 6, pp. 32 and 33).

Sometimes by using the amount of nitre calculated from a preliminary assay very little or no lead is reduced. This may happen when an ore contains much iron oxide, and some sulphur. The litharge combines

with the iron oxide, and some of the lead
unites with the sulphur. An addition of
nitre or a large addition of litharge will
remedy this.

Suppose the ore has an oxidizing power.
Take 5 grams of ore, 30 grams litharge, 18
grams sodium bicarbonate, 4 grams silica, 1
gram flour (or 2 grams argol), borax cover.
Proceed as directed above. One gram of
flour should reduce from 12 to 16 grams of
lead, if the ore is neutral. From the results
obtained, the weight of the reducing agent
that must be added to bring down a button
of the proper weight can be calculated.

Let us assume, for convenience of calculation :

1 assay ton...................... 30 grams
1 gram flour reduces .... 14 grams lead (12 to 16)
5 grams ore with 1 gram flour gave 11   "   "
   Hence, 5 grams ore oxidized
    (14 — 11) 3   "   "
1 assay ton (6 × 5 grams) will oxidize 18   "   "
   To neutralize the oxidizing ef-
    fect, it will take (18÷14).... 1.285 grams flour
   To bring down a button (23.8
    grams), it will take.......... 1.7 grams flour
   Or 1 assay ton of the ore requires 3   "   "   (2.985)

To a flux containing no reducing agent, for one assay ton of this ore must be added 3 grams flour. To a charge for one assay ton of the flux, page 4, must be added 1.2 grams flour, and silica.

The oxidizing and reducing powers of ores are different under different conditions. When the charge is acid, the sulphur oxidizes to SO_2 ; and, when much soda is present, the sulphur oxidizes to SO_3, and sulphates are formed. Under the latter condition more lead would be reduced.

A preliminary assay cannot be made for each ore in an office where much assaying is done.

ASSAY CHARGE FOR ORE CONTAINING SULPHUR.

Ore...	½ A. T.	
Litharge.....................................	30 grams	
Sodium bicarbonate.....................	24	"
Potassium carbonate....................	12	"
Silica................................. 1 to	10	"
Flour................................ 0 to	1	"
Cover of unfused borax..............	10	"
Nails...		

Desulphurization by Means of Iron Nails.

Ore containing a small percentage of sulphur can be assayed by adding 1, 2, or 3 20-penny iron nails to the above charge. Add sufficient silica to make a subsilicate with the bases present, making allowance for the amount in the ore. The more sulphur that is present, the less flour is needed.

If lime or magnesia are among the bases, add silica to make the slag given in Chapter V. The sulphur forms iron sulphide with the iron, a certain amount of which dissolves in the slag. If more is

formed than the slag can take up, a separate layer of matte is formed. If a matte forms, separate the lead button, grind up the matte and slag, mix with 1 assay ton litharge, 1 gram flour (or 2 grams argol), nails and silica as directed above, and use borax for a cover. Proceed as in a regular assay. In such an assay the slag would be very basic, without the addition of silica, and would attack the crucible by combining with the silica in the crucible.

Cupel this button with the first one obtained, or, if too large for cupellation, scorify the buttons to the proper size; or remove the matte from the slag, pulverize it, put it into a scorifier, and roast it "dead" (see under Roasting), then add the first button obtained, silica, test-lead, and borax-glass, and scorify as directed under 9.

If a button is obtained in any assay that is very large, it should be scorified to the proper size. There is less loss in scorification

than in cupellation; but, when the lead in a button is small in proportion to the silver, there is more loss of silver in scorification than in cupellation.

If the button is of the right size, but hard from the presence of iron, copper, or other base metals, or brittle from the presence of sulphur, arsenic, antimony, etc., scorify it with from 10 to 25 grams of test-lead and a little borax-glass.

Roasting.—Ores containing a large percentage of sulphur should be roasted. If the ore contains iron pyrites, put 1 assay ton of the ore into a clay roasting-dish, and perform the roasting in a muffle. Keep the temperature low at first or the ore will fuse and agglomerate. The rapid disengagement of volatile elements would also cause loss mechanically. Raise the temperature gradually, and stir the ore with a stout wire or iron rod, made by flattening one end, and bending it at a right angle, about an inch from the end. When,

on stirring, no more burning is seen, gradually raise the temperature to a dull red heat.
Then take the roasting-dish with the ore out
of the muffle, and allow the ore to cool, or
pour it on an iron plate to cool. When cold,
if 1 assay ton was taken, mix it with 42
grams litharge, 42 grams sodium bicarbonate,
from $\frac{1}{2}$ to $1\frac{1}{2}$ assay tons of silica, according to
the percentage of iron present; from 2 to 4
grams flour, according to the amount of sesquioxide of iron formed; and use about 15
grams unfused powdered borax for a cover
(or mix with from 8 to 10 grams borax-glass,
and use salt for a cover).

If all the sulphur has been burned out
("dead" roasted), and the iron is oxidized
to sesquioxide, it needs the reducing agent
prescribed. If not all the sulphur is burned
out, the ore may still have a reducing power;
and the addition of a reducing agent may
bring down too large a button.

When the percentage of sulphur is very

high, add an equal weight of silica to the ore before roasting, which will help to prevent the ore from agglomerating. Take the amount of silica added into consideration in fluxing the ore.

Ores can also be roasted over the furnace in a smooth, chalked, cast-iron pan, or in a crucible in the crucible-furnace. Proceed as directed above, and, if the roasting is done in a crucible, use the same crucible in which to make the fusion.

If sulphide ores are not roasted, oxysulphurets * may form, which are very fusible, but resist reduction at the temperature of the assay-furnace, and carry silver into the slag. In the fusion of ores containing arsenic and antimony, arseniates and antimoniates are formed, which carry silver into the slag. These ores should be roasted. Mix them with silica before roasting ; and, after roast-

---

* Ricketts and Miller's Notes on Assaying, pp. 94 and 95 ; Hiorns's Practical Metallurgy and Assaying, page 258.

ing, add some powdered charcoal to reduce arseniates and antimoniates, which may have been formed. Burn out all the charcoal.

When sulphates are formed during roast-ing that cannot be broken up at a dull red heat, mix the cool, roasted ore with some ammonium carbonate, cover, and heat until fumes cease. This converts the sulphates into ammonium sulphates, which are vola-tilized. Treat copper sulphide ore in this way.

Corrected Assays.—Slags from rich ores, or from ores containing much ferric oxide, Fe_2O_3, or slags containing many metallic oxides, should be assayed, and the button added to the first button. The slag from ores contain-ing zinc, arsenic, and antimony usually con-tain silver, if the ore contained silver. The cupels absorb silver and gold to some extent. Cupels in which large buttons of silver were cupelled should be assayed. Oxides of the base metals carry gold and silver into the

cupel. Copper oxide in particular carries much gold and silver into the cupel.

Assay Charges for Slags and Cupels.—Slag : If one assay ton of ore was taken, mix the pulverized slag with 1 assay ton litharge, 1 gram flour, silica, if basic, and use borax for a cover. A small amount of soda may be added. Charges should vary according to the character of the slag. Scorification slags generally need only a little flour or argol, and a little borax. *Cupels :* The phosphates present in the bone-ash make the slag from cupels pasty. Fluorspar or borax will make the slag fluid. For assaying a cupel in which a button of 20 grams or more has been cupelled, proceed as follows : Remove the saturated part of the cupel, pulverize it, mix with 50 grams litharge, 30 grams sodium bicarbonate, 30 grams borax-glass, and flour. The slag and the cupel from the same assay can be fused together, fluxed as follows : 50 grams litharge, 50 grams sodium bicarbonate, 1

gram flour or 2 grams argol, from 45 to 50 grams borax-glass; and an amount of silica according to the character of the slag.

General Crucible Charges.—From the fore-going discussion, the student will see that different ores require different fluxes. For silicious ores free from sulphur, etc., the following charge will answer:

```
Ore.........................   1 A. T.
Litharge ....................  42  grams
Sodium bicarbonate...........  42  grams
Flour .......................  1.7 grams
Cover............. 15 grams unfused borax
```

For ores containing only a small amount of sulphur add to the above charge one, two, or more 20-penny iron nails, and silica in proportion to the bases present, as ex-plained in Chapter V.

Roast ore containing much sulphur, arsenic, etc., add silica, and, if roasted "dead," add also from $\frac{1}{4}$ to 1 gram flour to the above formula, according to the amount of sesqui-oxide of iron present.

For almost pure iron pyrites the scorifica-tion method is usually preferred.

Ores with a gangue of lime or baryta need more soda, borax, and silica.

Mitchell's Formula:

Ore.............................. 1 A. T.
Sodium bicarbonate.............. 1 " "
Litharge .... 5 " "
Borax-glass.. .... 1 " "
Salt............................ Cover

And argol or nitre to bring down a button of the required size.

Aaron's General Formula:

Ore................... 1 A. T.
Litharge............. 1½ " "
Soda .... 3 " "
Borax............... ½ " "
Flour .... $\frac{1}{10}$ " "
Iron, 1 to 3 nails.
Salt................ cover

Melt and leave in strong fire about 20 minutes after fusion.

Some assayers reduce all the lead in the litharge used, and slag the gangue with borax,

soda, and silica. Aaron was the first to describe this method. Beginners usually do
not succeed very well with this method.

Special Methods.—For the assay of ores
and materials containing copper for gold and
silver, see Chapter IX.

The Assay of Galena for Silver.—It is the
general practice to make a scorification assay
for silver. The assay of galena for silver by
the crucible method, using the following
charge, gives good results :

Galena..................................	$\frac{1}{4}$ A. T.
Sodium bicarbonate...................	18 grams
Litharge.........	20 "
Borax (unfused) cover	18 "

Start at a low-red heat and finish at a
higher temperature.

Telluride Ore.—Assay telluride ore by the
crucible method. Assay charge :

Litharge................................	85 grams
Sodium bicarbonate...................	40 "
Silica...................................	10 "
Flour....................................	
Borax (fused) cover.	12 "

The amount of flour that it is necessary to add, if any, depends on the reducing power of the ore. If the ore is low in telluride, take $\frac{1}{2}$ an assay ton of the ore; and if it is high in telluride, take $\frac{1}{10}$ or $\frac{1}{20}$ assay ton of the ore, and take the full charge of the flux given above. Cupel the button, without scorifying it, at a temperature producing litharge crystals. The ore should be ground fine enough to pass a 120- to 150-mesh sieve; and the fusion, especially the first part of it, should be made at a low heat. Make a corrected assay (see page 70).

Tables of Crucible Charges.—Most practical assayers use a mixed flux and measure out the amount for each charge. The tables of crucible charges quoted on pp. 76, 77, and 78 are of value to give the student a better understanding of the relations of the various gangue materials and fluxes, and in particular cases they may be of value to treat the ore.

CRUCIBLE CHARGES. (Tabulated from Ricketts and Miller's Notes on Assaying).

Ore.	A. T. Ore.	A. T. Litharge.	A. T. Sodium Bicarb'nate.	Grams Argol.	Grams Nitre.	Grams Borax-glass.	Cover.	Remarks.
Quartz, rich........	1	2	1	1¼	...	10	salt	Same charges, with addition of from 1 to 3 A. T. of silica. If the gangue is magnesite or barite, increase both the soda and silica.
Quartz, low grade.	3	2	2	1¾	...	10	salt	
Quartz, tailings...	4	2	4	2	...	10	salt	
Pure ore, basic gangue....	...	...	...	...	...	...	...	
Galena	1	1	1	...	20	10	salt	Roast, and add silica, 2 A. T.
Arsenopyrite......	2	3	1¼	2¼	...	10	salt	Mix 1 A. T. ore with 1 A. T. clean sand and roast; then add a large spoonful of ground charcoal and heat until no more sparks are seen.
Antimony sulphide	...	2	1	2	...	10	salt	
Gray copper......	...	3	2	3¼	...	10	salt	As given for arsenopyrite. In either case the buttons may need rescorifying to remove impurities before cupellation.
Iron pyrite.....(1)	2	3	2	...	...	...	salt	Roast, and add 2 A. T. silica.
" " .....(2)	1	2	2	...	50	...	salt	Silica, 1½ A. T.
" " .....(3)	1	1	...	...	...	...	salt	Potassium cyanide, 2½ A. T. Moderate fire.
Copper pyrites	...	...	...	...	...	...	...	Roast with addition of ammonium carbonate and use charges given for iron pyrites. The buttons may require scorifying to remove copper.
Iron oxide.......	...	...	...	...	...	...	...	See charge (1) under pyrites.
Tellurides........	1½	4	1	1½	...	10	salt	Silica 1 A. T.
" low grade	2	4	2	2	...	10	salt	Silica 1½ A. T.

For ores (iron pyrite group): {For ores containing but little pyrites use one of the charges given for a pure ore, with the addition of a few nails or wires. This method of desulphurizing is not recommended for ores containing much sulphur or concentrates.

For tellurides: {Scorify the buttons if brittle. Cupel at a low heat near the end, as the button has a tendency to separate into small particles.

TABLE OF CRUCIBLE CHARGES.*

Ore.	Character of the Gangue.	A. T. Ore.	Gms. of Lead Flux.†	Gms. of Soda Bicarbonate.	Gms. of Litharge.	Gms. Potassium Ferrocyanide.	Gms. Nitre.	Gms. Silica.	Gms. Argol.	Loops of Iron Wire.	Gms. Borax-glass.	Cover.	Remarks.
Oxidized......	Neutral, no Pb...	½	30		25							borax	If cover of salt is used in place of borax, add 3 to 5 gms. borax-glass.
Quartz........	No bases...........	½			75				2			borax	Special method. If oxide iron is present, add soda in proportion.
Quartz........	No bases........	½	30	30	20			15				salt	
Oxidized.	Basic, no Pb	½	30–40		20							borax	If gangue is oxide or carbonate of iron, add 2 to 3 gms. of argol.
Oxidized......	Basic, with BaSO₄	½	40	20	25			15		2		borax	Borax-glass may be substituted for some of the silica.
Galena........	Lead 84 per cent..	½	20			10						salt	Heat gradually until mass subsides.
Galena........	Silicious Pb 42 per cent.....	½	15	20	20		5					salt	Litharge is added according to the lead contents of the ore.

* Furman's Manual of Assaying.   † Lead flux, p. 46, (1) under 1.

TABLE OF CRUCIBLE CHARGES—(Continued).

Ore.	Character of the Gangue.	A. T. Ore.	Gms. of Lead Flux.	Gms. of Soda Bicarbonate.	Gms. of Litharge.	Gms. Potassium Ferrocyanide.	Gms. Nitre.	Gms. Silica.	Gms. Argol.	Loops of Iron Wire.	Gms. Borax-glass.	Cover.	Remarks.
Lead carbonate	Neutral Pb 40 per cent......	½	30	10	15	...	...	...	...	...	...	borax	Litharge added according to the lead contents of the ore.
Iron pyrites....	None........	½	...	35	20	...	5	15	...	6	...	borax	Collect matte, if any, and scorify with lead button.
Copper pyrites	Iron pyrites	½	...	35	30	...	5	15	...	6	...	borax	Collect matte and scorify with lead button. If button is hard, add test-lead.
Lead matte....	....	½	15	30	20	...	5	15	...	5	...	borax	Special method is preferable.
Copper matte .	....	½	15	30	35	...	5	15	...	5	...	borax	If button is hard or brittle, scorify with lead. Scorification preferable.
Tellurides.	Silicious.....	½	30	30	40–80	...	...	...	...	...	...	salt	Special method.
Tellurides.....	Silicious.....	½	...	...	80	17	...	...	2	...	...	salt	Special method.
Arsenical.....	....	½	...	15	30	...	...	...	...	...	...	salt	Scorify button. Scorification assay is preferable.
Slags.....	....	1	20	40	10	...	...	...	...	...	10	...	If slag contains matte, add loop of iron wire.

SCORIFICATION CHARGES.

1/10 A. T. Ore.	Grams of Test-lead.	Grams of Borax-glass.	Remarks.
Quartz.........	25–30	0.00–0.10	
Copper ores.....	35–50	0.30–0.50	Add 0.5 to 1.5 grams silica. Low temperature. If button contains copper, rescorify with lead, silica, and borax-glass.
Iron pyrites.....	30–40	0.50–0.80	Add 1 to 1.2 grams silica. Mix the ore with 5 to 10 grams litharge and cover with the test-lead.
Iron oxide......	25–30	0.50–0.80	Add 1 to 2 grams silica.
Galena.........	16–18	0.50–0.60	
Lime........ ⎫			
Magnesia..... ⎬	30	0.50–2.00	Add 1.50 grams silica, if gangue is lime or magnesia. If gangue is baryta, add 2.25 grams silica. Sodium carbonate helps assay.
Baryta....... ⎭			
Chloride.	25	0.10	Low heat until covered, then raise temperature.
Zinc blende....	35–45	0.50–0.60	High temperature.
Arsenic....... ⎱	45–55	0.50–1.5	Mix the ore with 5 to 10 grams litharge and cover with the test-lead. May need several rescorifications. Powdered charcoal aids fusion.
Antimony.... ⎰			
Zinc precipitates.	35–50	0.30–0.60 Add a little at a time until the s l a g i s fluid.	Mix the zinc precipitates with 10 grams litharge and cover with the test-lead.

SCORIFICATION CHARGES.*

Ore $\frac{1}{10}$ A. T.	Grams of Test-lead.	Grams of Borax-glass.	Remarks.
Galena....	15–18	up to 0.5	
Galena with blende and pyrite	20–35	0.4–0.8	
Iron pyrite.............	30–45	0.3–0.8	
Arsenical pyrite........	45–50	0.3–1.5	High temperature. Addition of litharge helps assay.
Gray copper....... ...	35–48	0.3–0.5	Low temperature.
Blende.................	30–45	0.3–0.6	High temperature. Addition of oxide of iron helps assay.
Copper ores and mattes	35–40	0.3–0.5	Low temperature. If necessary, the button should be rescorified with lead.
Lead mattes.	25–35	0.5–1.0	
Furnace accretions.....	25–50	0.3–1.5	
Tellurides	50	0.3	Add a cover of litharge and rescorify the button.
Silicious	25–30		
Basic.................	25–30	0.5–2.0	If the ore contains much lime or magnesia the addition of sodium carbonate helps the assay.
Basic with barium sulphate.......	25–30	0.5–1.5	Addition of sodium carbonate helps assay.
Lead carbonate........	10–15	up to 0.5	
Speisse................	30–60	0.3–0.5	High temperature. Rescorify the button with lead if necessary.

* Furman's Manual of Assaying.

INFLUENCES OF BASE METALS ON CUPELLATION.

	Scoria.*	Stains Cupel.	Remarks.
Iron	Brown.	Brown.	The oxide is not easily fusible with lead oxide. If much iron is present, the button may "freeze." Add lead and raise the temperature. In general, these remarks apply also to tin, nickel, cobalt, copper, and platinum.
Tin	Gray.	Gray.	See remarks on iron.
Copper		Green to blackish green.	See remarks on iron. Also carries much gold and silver into the cupel. Usually not completely removed from the bead. When much copper is present, a rose-colored coat of copper oxide forms on the outside of the cupel.
Zinc	Pale yellow.	Pale green.	Button is slightly crystalline. Burns with a blue flame, and volatilizes, carrying gold and silver with it.
Nickel and cobalt	Dark green.	Green.	Button is crystalline.
Manganese	Dark.	Black.	Corrodes the cupel.
Arsenic	Yellowish white.		Spirts. Volatilizes, carrying gold and silver with it. Remove by scorification, if much is present.
Antimony	Yellow.		Spirts. Volatilizes, carrying gold and silver with it. Remove by scorification, if much is present. Causes the cupel to crack.
Aluminum	Gray.		Retards cupellation.
Chromium	Brick-red.	Brick-red.	Retards cupellation.
Tellurium		Yellow.	Causes loss by volatilization, and causes subdivision of the bead.
Titanium		Deep orange.	

* Scoria may entangle lead, and thus occasion loss, especially if the cupel is disturbed in such a way as to bring lead in contact with the scoria.

CHAPTER VII.

AMALGAMATION AND CHLORINATION TESTS.

Amalgamation and chlorination tests are usually intended to give an approximate indication of the percentage of gold and silver that can be extracted by these methods on a working scale.

Amalgamation Test. —Pulverize about 5 lbs. of ore fine enough to pass through a 40-mesh sieve. Sample it down to about 1 lb. (see 3 ; and 1 under 12). Pulverize this sample fine enough to pass through an 80-mesh sieve, and assay it in the regular way. Mix the remaining ore and weigh out 3 lbs. and put $\frac{1}{4}$ lb. of it into each of the twelve bottles on the shaking-frame (Fig. 37). Add sufficient water to

each bottle to make the ore of the consistency of very thin mud. Add about $\frac{1}{2}$ oz.

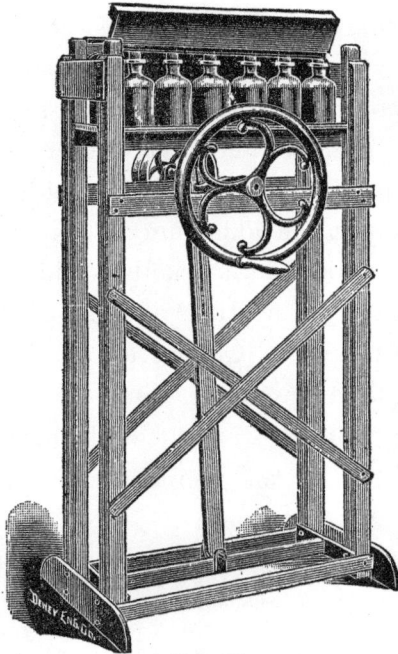

FIG. 37.

of clean mercury to the contents of each bottle, and run the frame at a high speed for

an hour or more. Then empty the bottles into a miner's pan, and wash the tailings into another pan or tub. Pan out the mercury and the concentrates (particles of gold that will not amalgamate, sulphurets, etc.). Dry the concentrates, weigh, and assay them. Dry the tailings and assay them. Retort the mercury, or squeeze it through a clean buckskin, put the amalgam left in the buckskin into a porcelain crucible, and drive off the mercury by heating at first gently and finally to redness. Collect the gold, wrap it in lead-foil and cupel. Small amounts of amalgam can be treated with nitric acid, which dissolves the mercury and leaves the gold.

Ore assayed ... . 3 ozs. gold per ton
Amount of ore taken for amalgamation.. 3 lbs
Assay value for 3 lbs. of this ore........ 2.16 grains gold *
Gold extracted by amalgation from 3 lbs. 1.36 grains gold
This corresponds to a yield of........... 63 per cent of
assay value
Concentrates from 3 lbs.............. 3 ozs

* See tables on page 142.

One ton would give. .... 125 lbs
Sixteen tons of ore would give.......... 1 ton of concen-
trates
Concentrates assayed per ton concentrates 16.48 ozs.
The 125 lbs. in a ton of ore would give... 1.03 ozs.
Tailings assayed........................ ... 0.09 oz.

The silver can be calculated in the same way.

Retorting. — The mercury is usually squeezed through a buckskin and the amalgam put into a retort (Fig. 38), which has

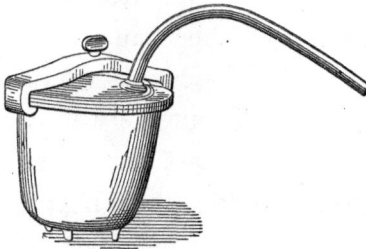

FIG. 38.

been rubbed with chalk on the inside. The amalgam should not be put into the retort in lumps, nor pressed down. Lute the lid on with a paste of flour, and fasten the clamp. Do not fill the retort more than about three-

fourths full. Apply a low heat at the top and gradually increase the heat to redness. After no more mercury comes over, increase the temperature to a cherry-red heat. During the heating, the condensing-pipe must be kept cool by keeping water running over it. The end of the pipe is usually placed in water. If the condensing-pipe sucks water at any time, it must be immediately lifted out of the water to prevent the water from being sucked into the retort and thus cause an explosion. It is safer to attach a canvas or rubber bag to the end of the pipe in the water.

An amalgamation test can be made on a smaller scale by shaking, in a bottle, a pound of ore with an ounce of mercury. The preparation of the ore and the other operations are the same as already explained.

Chlorination Test.—Sample and assay the ore as directed under amalgamation assay. Weigh out one pound of the ore and roast it. After it is cold, moisten the ore

sufficiently to make it cohere slightly when compressed in the hand. Put it into a wide-mouth bottle that has an opening at the bottom. The bottle should not be over half full. Pass a stream of chlorine into the bottle through the opening at the bottom. When chlorine begins to escape at the top, close the bottle and pass the chlorine a while longer. Then close the lower opening, and let it stand twenty-four hours. If at the end of that time, chlorine is still in excess, leach the ore with hot water until the filtrate gives no reaction for chlorine. Dry the residue and assay it. From the difference between this and the first assay, the approximate percentage of extraction can be calculated. The gold in the solution can be precipitated with ferrous sulphate. Collect the precipitate and cupel.

CHAPTER VIII.

THE ASSAY OF ORE CONTAINING COARSE METAL.

WHEN ore contains scales of gold or other coarse metal that will not pass the sieve, it may be assayed by one of the following methods :

1. Weigh (in grams) the pulp that passes the sieve, and determine its value by crucible or scorification assay in the regular way. Weigh (in grams) also the scales or other coarse metal, and determine its value in the same way. If the scales consist of gold and silver only, wrap them (or an equal part of them) in lead-foil, and cupel them in the regular way. If the gold only is to be determined, add the necessary silver to insure separation (see pp. 21 and 37) before cupelling the scales.

Example.—Suppose the pulp that passed the sieve weighed 112.664 grams, and the coarse metal weighed 4 grams. 1 assay ton of the pulp yielded 0.002 gram gold. 112.664 ÷ 29.166 (grams in one assay ton) = 3.86 assay tons in the pulp that passed the sieve. If 1 assay ton gave 0.002 gram gold, 3.86 assay tons would give 3.86 × 0.002 gram = 0.00772 gram gold. The 4 grams coarse metal gave 0.014 gram gold. 0.00772 + 0.014 = 0.02172 gram gold in the whole sample.

112.664 grams that passed the sieve + 4 grams coarse metal = 116.664 grams, the weight of the whole sample.

116.664 ÷ 29.166 = 4 assay tons in the whole sample, which yielded 0.02172 gram gold. 1 assay ton (0.02172 ÷ 4) will give 0.00543 gram, which = 5.43 ozs. Troy to 1 ton ore of 2000 lbs. Avoirdupois (see assay ton weights, p. 9). The silver can be calculated in the same way.

2. If the coarse metal consists of copper or silver, dissolve it in nitric acid; if it consists of gold, dissolve it in aqua regia ($3HCl + HNO_3$). Evaporate the solution to a small bulk, and add it to the ore that passed the sieve in such a way that it will not run or soak through to the bottom. Dry the sample at a temperature not above 100° C. Then pulverize it again, pass it through the sieve, and mix it *thoroughly.* Assay it in the regular way.

If copper is present, assay it by one of the methods given in Chapter IX.

3. Place some of the pulverized ore on the bucking-board, or in a grinder, add the scales, and grind hard until the scales pass the sieve. Mix the ore thoroughly, and assay it in the regular way.

By this method it is difficult to get assays that agree closely.

CHAPTER IX.

THE ASSAY OF COPPER ORES AND COPPER-BEAR-ING MATERIALS FOR GOLD AND SILVER.

THE operations in making these assays are not described in detail as the student, if he has mastered the preceding chapters of this book, is familiar with the operations of the crucible and scorification assays. The crucible and scorification charges for the different gangue materials are also given on preceding pages.

Ores and materials containing copper may be assayed by one of the following methods:

1. The gold can be determined by crucible assay, if the percentage of copper is not above about 15 per cent.

Take $\frac{1}{2}$ A. T. ore, or $\frac{1}{4}$ A. T., if the percentage of copper is high. Add silica, and, if

sulphur is present, roast the ore, cool, add ammonium carbonate, and heat (see roasting, and crucible charges for ores containing copper). Scorify the button with silica, test-lead, and borax to remove copper, and cupel.

If the ore is low grade, make a number of assays, and scorify the resulting buttons to remove copper, and to reduce them to the proper size for cupelling.

Determine the silver by scorification assay. If the ore is low grade, make a number of assays, and scorify the buttons as directed above (see scorification charges for copper ore).

2. Ores and mattes containing much more than 15 per cent copper can be assayed by the scorification method (see scorification charges for ores and mattes containing copper). If the percentage of copper is high, take $\frac{1}{20}$ A. T., and make from 3 to 10 scorifications. Scorify the resulting buttons with

a little silica, test-lead (if they contain much copper, or if the amount of lead is small), and a little borax.

3. The combination wet-and-dry method is, in general, as follows: 1 A. T. is treated with nitric acid until all the copper is dissolved. The red fumes are expelled by boiling, after which the solution is filtered. If ore is treated, the residue is brought on to the filter, washed, and dried. Take the residue from the filter, burn the filter-paper in a porcelain crucible, and add the ash to the residue, flux and assay as usual. Add sufficient normal salt (NaCl) solution to the filtrate to precipitate the silver. Avoid an excess, as silver chloride is soluble in salt solution. Allow the silver chloride to settle, then filter, wash the silver chloride on to the filter, put the filter-paper with the silver chloride into a scorifier, burn the filter in front of the muffle at a low temperature, add test-lead, scorify, and cupel.

It is, of course, evident that, if ore or other materials are assayed in which there is gold that is alloyed with less silver than will insure separation (see pp. 21 and 87), not all the silver would dissolve, and the button obtained from the silver chloride would not be the whole amount of the silver in the material. In this case the filter-papers may be burned, and, together with the dry residue and the silver chloride, fluxed (and placed in the same crucible) and assayed as usual.

If copper borings are assayed, the residue, silver chloride, and both filter-papers can be put into the same scorifier (the papers burned) and scorified.

If the silver chloride is free from foreign materials, as sand, etc., it may be cupelled directly (omitting the preliminary scorification). The silver may be precipitated by bromides (see (1), below).

As soon as the filter has drained, dust about 2 grams powdered lead over the inside

of the wet filter containing the silver chloride, fold it so that several thicknesses of the filter will envelop the silver chloride, bring a red-hot cupel in front of the muffle, put about 4 grams test-lead into it, then put the filter into it, and allow it to burn in front of the muffle. Then cover the ashes with about 2 grams test-lead, put the cupel carefully into the muffle, and cupel at a temperature producing litharge crystals.

If the gold is also to be determined, and it is free from foreign materials, the filter containing it may be burned in a coil of platinum wire and the ashes allowed to drop into the filter containing the silver chloride ; then fold the filter and proceed as directed above.

After the solution of the copper borings is complete, and the red fumes are boiled off, add a drop of normal salt (NaCl) solution before filtering off the gold. The silver chloride formed will collect the fine particles of gold.

There are various modifications of this method:

(1) *Whitehead's Method.**—Dissolve 1 to 4 A. T. in a large beaker (500 c.c. capacity) by the gradual addition of strong nitric acid; drive off red fumes by heating in sand-bath; add 50 c.c. saturated solution of lead acetate; stir; add 1 c.c. dilute sulphuric acid, and allow lead sulphate to settle. Filter; wash with cold water, dry in scorifier; burn filter-paper; scorify with test-lead; cupel, weigh, and part as usual.

Dilute the filtrate to 1000 c.c.; divide in halves of exactly 500 c.c.; add to each saturated solution of sodium bromide so long as a precipitate forms. A large precipitate of lead bromide collects and envelopes the silver bromide permitting immediate filtering without loss. Filter; wash with cold water; dry filters and precipitates; brush into small crucibles; mix each with three times its

---

* Transactions of the American Institute of Mining Engineers, March, 1895.

weight of carbonate of soda and some flour or argol as reducing agent; cover with borax-glass; fuse for lead buttons; cupel and weigh. The two results should agree closely.

(2) *Godshall's Method.**—Dissolve 1 A. T. of copper borings in dilute nitric acid (90 c.c. strong acid to 100 c.c. water). Evaporate the solution to expel free nitric acid, add 20 c.c. sulphuric acid, and continue the evaporation. Add hot water to take up the copper salts, dilute to about 800 c.c., and allow to cool. Then pass a rapid stream of sulphuretted hydrogen into the solution for about two minutes. Stir the liquid containing the precipitate, then allow the precipitate to settle about half an hour, then filter rapidly.

Dry the copper sulphide containing the gold and silver, burn the filter in a scorifier, and scorify the residue with 50 grams test-lead.

---

* Transactions of the American Institute of Mining Engineers, February, 1900.

The statement has been made that, in parting, loss of gold results from the action of
nitrous acid. It has also been stated that
in the wet-and-dry assay of copper-bearing
materials, loss of gold results from the combined action of nitric acid and nitrous acid in
a hot solution. Bulletin No. 253, U. S.
Geol. Surv., states: "The results of these
severe tests show that solvent action by nitrous acid during an assay need not be considered."

For the determination of gold in copper-
bearing materials, the scorification method
(see 2, above) gives better results.

(3) Mr. W. Randolph Van Liew has
evolved a method* by which he claims an
average gain of 6.75 per cent of gold in
favor of his method, as compared with the

---

* The *Engineering and Mining Journal*, April 21st and
28th, 1900.

usual wet-and-dry method. His method is substantially as follows:

Take 1 A. T. copper, treat with 350 c.c. very cold water, and 100 c.c. nitric acid (sp. gr. 1.42), and set the beaker in a cool place (temperature should be kept down to 15° or 16° C.). At the end of 18 or 20 hours, add sufficient nitric acid to dissolve the remaining copper (amount of acid will vary from 0 c.c. to 30 c.c. nitric acid, sp. gr. 1.42). At the end of 24 to 26 hours, the solution of copper is complete. Remove the oxides of nitrogen by compressed air. Insert the pointed end of a pointed glass tube into the solution through which pass air into the solution, by means of a blower or other apparatus, until the red fumes are removed, which takes from 20 to 30 minutes. No heat is applied at any stage of the process.

Add to the cold solution a slight excess (of from 2 c.c. to 4 c.c.) of normal NaCl solution. Experiments have shown that no

difference was made whether the gold was filtered off before or after the addition of normal NaCl solution.

Allow to stand over night, then filter, wash the entire contents to the point of the filter-paper, cover with from 4 to 6 grams test-lead, and, after the filter-paper has drained, place it in a $2\frac{1}{2}$-inch scorifier, in the bottom of which there is about 1 gram test-lead.

Dry and burn the paper in a muffle at a temperature below incipient redness. At the end of the yellow flame of the paper, remove the scorifier, and allow the charred paper to burn outside the furnace. By this slow combustion at a low temperature, loss of silver, by its reduction from silver chloride, is avoided.

At the end of about 20 minutes the paper will have burned. Now add from 3 to 4 grams litharge, and from 3 to 4 grams borax-glass.

If all the copper has been washed out, and there are no impurities, no scorification is necessary. The operation is simply to melt and collect the gold and silver, after which the scorifier may be poured. Cupel the button at a temperature giving heavy litharge feathers, and allow to blick at the same temperature.

The time of operation is 48 hours.

Mr. Van Liew reported 6 tests, by this method, which showed a loss of silver of from 1.2 to 2.2 per cent, or an average loss of silver of 1.7 per cent. Eighteen tests on c.p. gold showed a loss of gold of 0.00 per cent to 0.50 per cent, or an average loss of gold of 0.13 per cent.

CHAPTER X.

THE DRY ASSAY OF MERCURY.

THE dry assay of mercury is effected in various ways: (1) The ore is pulverized and mixed with reducing agents, and then put into a retort, combustion-tube, or other distilling apparatus. The mercury is then driven off by heat, condensed, collected, and weighed. If the ore is rich and a large amount of it is taken, this method may give approximately accurate results. (2) Better results are obtained by receiving the mercury on gold-foil with which the mercury amalgamates, but gold is expensive. After each assay, the mercury must be driven out of the gold-foil, which usually occasions a loss of gold. After a number of determinations,

the gold must be remelted and rolled out again.

On account of these and other objections to the use of gold, Mr. Richard E. Chism employs silver as a recipient for the mercury. The following is condensed, with some changes, from his article * describing his method :

Apparatus † and Flux.

Heating Apparatus.—A glass alcohol-lamp can be used.

Retort or Crucible.—Use a glazed white clay (or porcelain) crucible, in the form of a truncated cone, about 2 centimeters outside diameter at the bottom, and 3.5 centimeters diameter at the mouth; height about 4.5 cen-

---

* Transactions of the American Institute of Mining Engineers, October, 1898.

Mr. Chism claims as original the use of silver for receiving mercury (though this was suggested before), and the use of a separate vessel to cool the receiving surface by contact.

† See Fig. 39.

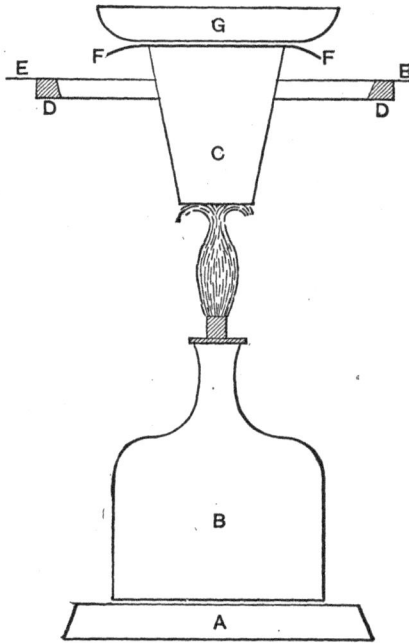

FIG. 39.—APPARATUS FOR MERCURY-ASSAY ONE-HALF
ACTUAL SIZE.

A, base of retort-stand. *B*, spirit-lamp. *C*, retort or annealing-cup. *D D*, retort-stand ring, which serves as sup-
port to the apparatus. *E E*, tin shields. *F F*, silver-foil
for receiving the mercury. *G*, cooling-cup.

timeters. The mouth of the crucible should
have an even surface.

Shield.—To prevent the direct heating of
the upper part of the crucible and silver-foil,
use a circular tin shield about 13 centimeters
in diameter, with a hole in the center large
enough to pass the crucible partly through,
leaving about 1 centimeter of the crucible
above the shield.

The Recipient.—Use a piece of pure silver-
foil (rolled silver) about 5 centimeters square,
and about 0.02 millimeters in thickness, on
which to receive the mercury. It should be
large enough to cover the crucible and leave
a margin all round of about one-half centimeter.

Cooling-cup.—For cooling the silver-foil,
use a silver dish of a wide pattern like an
evaporating-dish. Silver is a good heat-con-
ductor. A copper dish could be used. The
bottom of the dish should be a little larger
than the mouth of the crucible. Keep the

bottom of the dish polished to enable you to discover any mercury that might soak through the silver-foil. Should this happen, drive off the mercury from the dish by heat, and repeat the assay with new silver-foil and less ore.

Flux.—Use iron filings, the finer the better. They should pass a 60-mesh sieve. Remove most of the grease with strong alcohol, and then heat them to redness for some time in a muffle. Then keep the filings in a glass bottle with a rubber stopper.

To Make the Assay.

Take from one-half to one gram of the ore, prepared as directed under 3. If the ore is very rich, take less. Thoroughly mix the ore in the crucible with 5 grams of the prepared iron filings, and put about one gram of the iron filings on top of the charge as a cover. Now hang the *crucible*, by its tin shield, from the ring of a ring-stand. Carefully smooth the *silver-foil* (see The Recipient above), and

ignite it in the flame of an alcohol-lamp.
Care must be taken not to overheat the foil,
or it will fuse.   Cool the foil in a desiccator
(Fig. 40), and then weigh it accurately on an

FIG. 40.

analytical balance.   Press it gently on the
mouth of the crucible until it assumes the
shape of the mouth of the crucible.

Place the *cooling-cup* upon the silver-foil
on top of the crucible, and fill the cooling-cup
with cold water (ice-water, if at hand).

Place the *alcohol-lamp* under the crucible,
and arrange it to give a flame about 4 centimeters high, which shall barely spread out
at its point over the central part of the bottom of the crucible.

Continue the heating from 10 to 15 minutes. Ten minutes is too short for most ores, and anything over 15 minutes is apt to lead to loss of mercury.

If ice-water is not used, it may be necessary to renew the water in the cooling-cup once or twice during the heating.

When the heating is at an end, allow the crucible and contents to cool at least five minutes. When the silver-foil is removed, a distinct mercurial stain will be seen upon its lower surface, if there was the slightest trace of mercury in the ore.

Convey the foil (under cover to avoid dust) to the balance, and weigh it. The increase in weight of the foil shows the amount of mercury on the foil.

In order to check the first determination, and make sure that all the mercury has been collected, place the silver-foil on the crucible again, and heat the same charge about 10 minutes more, allow to cool, and weigh again,

If the weight is constant, or if there is a slight decrease in weight, the amount of mercury obtained by the first weighing may be considered correct. If more mercury has been collected on the second weighing, repeat the determination with a new charge, and heat a longer time—five or ten minutes longer than the first time.

The foil can be preserved for future reference, or the mercury can be driven off by carefully heating it in an alcohol flame; then the foil can be used for another determination.

Polish the foil, if it is not bright, and carefully anneal, cool, and weigh it just before making each assay, as directed above.

Failures.—Failures may arise from too high or too long heating, from the foil being badly adjusted to the mouth of the crucible, or from the cooling apparatus not being properly managed.

CHAPTER XI.

MR. ALBERT H. Low, assayer and analytical chemist, of Denver, Colorado, has modified the cyanide method and the iodide method for the determination of copper. He has sent me the following descriptions of these methods as modified by him:

The Cyanide Assay for Copper.

Standardization of the Solution. — The standard solution should contain about 21 grams of pure potassium cyanide per 1000 c.c. Determine the exact standard as follows:

Dissolve about 0.200 gram, accurately weighed, of pure copper-foil in 5 c.c. of strong nitric acid. Use a flask of about 250

108

c.c. capacity. Without troubling to boil off
the red fumes, add about 80 c.c. of water and
10 c.c. of strong ammonia water. Cool the
mixture to the ordinary temperature. Titrate
with the cyanide solution, slowly and cautiously, so as to allow sufficient time for the
fading of color due to each addition. When
the blue color has become perceptibly weaker,
but is still fairly strong, dilute the solution so
that the final bulk will be about 150 c.c. Now
finish the titration by adding the cyanide in a
slow and regular manner, finally one drop at
a time until the blue tint is entirely discharged.
The exact end-point is best observed by the
aid of a vertical white background. From
the amount of cyanide required, calculate the
copper value per c.c.

The accuracy of all subsequent work with
the standard solution depends upon the operator's ability to duplicate the essential conditions of the final additions of cyanide. These
are temperature, bulk of solution, and speed

of working. Up to the point where the amount of cyanide added is insufficient to nearly discharge the color on long standing, the manner of adding it appears of no consequence; and the assay that has thus stood may be resumed and finished without detriment. The reaction proceeds rather slowly, and towards the end its speed is usually exceeded by that of the operator. It is therefore necessary, in finishing the titration, to proceed in a deliberate, methodical manner that can be duplicated in all subsequent work.

Assay of Ores, etc.—Treat 0.5 gram in a flask of about 250 c.c. capacity with about 6 c.c. of strong nitric acid. Boil gently nearly to dryness, and then add 5 c.c. of strong hydrochloric acid, and again heat until all soluble matter is taken up. Now add 5 c.c. of strong sulphuric acid, and boil until the white fumes are freely evolved.

Time may be saved and bumping avoided

by manipulating the flask in a holder over a naked flame.

Allow to cool, add 20 c.c. of cold water, and heat to boiling. When an ore contains much iron, an insoluble anhydrous sulphate is apt to be present at this stage, which will only slowly dissolve in the warm dilute acid, meanwhile remaining more or less in suspension with a milky appearance. This anhydrous sulphate retains copper. Do not proceed to filter, therefore, but keep the mixture warm, with occasional agitation, until the liquid clears and the residue appears normal. This may take several minutes. Finally filter, wash with cold water, and collect filtrate in a beaker about two and one-half inches in diameter. There should be about 75 c.c. of combined filtrate and washings.

Place in the beaker a piece of stout sheet aluminum about five and one-half inches long and five-eighths of an inch wide, bent into a triangle so as to stand on edge. The same

triangle will last for many assays. Add one drop of a mixture of equal parts of strong hydrochloric acid and water, cover the beaker, and heat to boiling.

A strong action is liable to occur unexpectedly, and it is usually necessary to lower the heat when boiling begins. Boil gently for 7 to 10 minutes, and then remove from the heat, and wash down the cover and sides of the beaker with cold water. Now add 15 c.c. of strong sulphuretted hydrogen water. This should produce little or no discoloration. Even though the copper be entirely precipitated, there is still danger of oxidation and loss during the subsequent washing unless sulphuretted hydrogen water be employed.

Pour the solution through a 9 c.m. filter, and rinse on the copper with sulphuretted hydrogen water. Wash once or twice with the same water, and then finish with pure cold water.

The metallic copper on the filter is more or

less mixed with copper sulphide, and there may be a little copper left adhering to the aluminum in the beaker. Pour over the latter 5 c.c. of strong nitric acid, and then wash this into a second beaker, using not over 5 or 10 c.c. of water. Now transfer the filter and copper into the dilute acid, and warm gently until all is dissolved, and the separated sulphur appears clean.

If the ore contains silver, it should be precipitated at this stage by the addition of a single drop of strong hydrochloric acid before the heating. Filter the copper solution into the original flask. Wash the filter thoroughly, but avoid getting the filtrate too bulky. Add 10 c.c. of strong ammonia water to the filtrate, cool to the ordinary temperature, and titrate with the standard cyanide solution precisely as in the standardization. Dilute with water towards the end, if necessary, so as to obtain a final bulk of about 150 c.c.

Should the presence of lead or other impur-

ity cause a milkiness in the blue solution, it is best to filter the mixture, when nearing the end of the titration, through a coarse rapid-running filter. From the number of c.c. of standard cyanide required, calculate the percentage of copper in the ore.

When the amount of silver is known, it need not be removed, but may be allowed for on the basis that $2Ag = Cu$. 100 ounces of silver per ton of 2000 lbs. will approximately equal 0.10 per cent of copper. Deduct, accordingly, 0.10 per cent of copper for every 100 ounces of silver per ton.

Copper Assay by the Iodide Method.

Prepare a solution of sodium hyposulphite containing about nineteen grams of the pure crystals to the liter. Standardize as follows : Weigh accurately about 0.200 gram of pure copper-foil and place in a flask of about 250 c.c. capacity. Dissolve in five c.c. of a mixture of equal volumes of strong

nitric acid (1.42 sp. gr.) and water; and then dilute to about fifty c.c., and boil until the red fumes are thoroughly expelled. This last is a very essential point. Remove from the heat and add a slight excess of ammonia water to the hot liquid. Ordinarily it suffices to add five c.c. of strong ammonia (0.90 sp. gr.). Now add acetic acid in slight excess,—say three c.c. of the 80 per cent acid in all. Cool to ordinary temperature and add three grams of potassium iodide, or five c.c. of a solution containing sixty grams of potassium iodide in 100 c.c. Cuprous iodide will be precipitated, and iodine liberated according to the reaction $2(Cu.2C_2H_3O_2) + 4KI = Cu_2I_2 + 4(K.C_2H_3O_2) + 2I$.

The free iodine colors the mixture brown. Titrate at once with the hyposulphite solution until the brown tinge has become weak, and then add sufficient starch liquor to produce a marked blue coloration. Continue the titration cautiously until the blue tinge

has entirely vanished. When almost at the end, allow a little time after the addition of each drop to avoid passing the point. One c.c. of the hyposulphite solution will be found to correspond to about 0.005 gram of copper. In the assaying of ores, when half a gram is taken, one c.c. of the standard hyposulphite would then equal about one per cent of copper. The reaction between the hyposulphite and iodine is $2(Na_2S_2O_3) + 2I = 2NaI + Na_2S_4O_6$. Sodium iodide and tetrathionate are formed.

The starch liquor may be made by boiling about half a gram of starch with a little water, and diluting with hot water to about 250 c.c. It should be used cold, and must be prepared frequently, as it does not keep well.

The hyposulphite solution made from the pure crystals and distilled water appears to be quite stable, showing little or no variation in a month, if kept under reasonable conditions.

Treatment of Ores.

To half a gram of the ore in a flask of 250 c.c. capacity, add about six c.c. of strong nitric acid, and boil gently nearly to dryness. Then add five c.c. of strong hydrochloric acid and again heat. As soon as the incrusted matter has dissolved, add five c.c. of strong sulphuric acid, and boil until the more volatile acids are expelled, and the fumes of sulphuric acid are coming off freely. This is best done by manipulating the flask in a holder over a naked flame. Allow to cool and add 20 c.c. of cold water, and heat the mixture to boiling. Allow to stand, hot, until any anhydrous sulphate of iron is dissolved, and then filter to remove more especially any lead sulphate. Receive the filtrate in a beaker about two and a half inches in diameter. Wash flask and filter with either hot or cold water and make the volume of the filtrate about seventy-five c.c. Place in

the beaker a piece of sheet aluminum prepared as follows : Cut a strip of stout sheet
aluminum five-eighths of an inch wide and
about five and one-half inches long, and bend
this into a ring so that it will stand upon
its edge in the beaker. The same aluminum
may be used repeatedly, as it is but little
attacked each time. Add one drop of a mixture of equal parts of strong hydrochloric
acid and water, cover the beaker and heat to
boiling. Allow to boil seven minutes, which
will be sufficient to precipitate all the copper
in any case, provided the bulk of the solution does not much exceed seventy-five c.c.
The aluminum should now appear clean, the
precipitated copper being detached or only
loosely adhering. Remove from the heat
and wash down the cover and sides of the
beaker with cold water. There is danger of
the finely divided copper being slightly oxidized and dissolved during the subsequent
washing. To prevent this, add at once

fifteen c.c. of strong hydrogen-sulphide
water. If the amount of the precipitated
copper is large, it is best to wash it by
decantation, as will be subsequently described; but, for quantities not exceeding
say 20 per cent, it is more convenient to proceed as follows: Pour the clear liquid
through a nine cm. filter and then wash on
the copper with cold $\frac{1}{2}$ s. water. The beaker
and aluminum, which may still retain some
adhering particles of copper, are now set
aside temporarily. Wash the copper on the
filter several times with cold water, and then
place the original flask under the funnel.
Now pour over the aluminum in the beaker
five c.c. of a mixture of equal volumes of
strong nitric acid (1.42 sp. gr.) and water,
and heat to boiling. Do not prolong the
boiling, or the aluminum will be unneces-
sarily attacked. Pour the hot acid very
slowly over the copper on the filter so as
to dissolve it all, and then wash beaker and

filter several times. Heat the solution in
the flask to boiling and thoroughly boil off
the red fumes; then replace the flask under
the funnel and pour five c.c. of strong bromine water through the filter. The bromine cleanses the separated sulphur left on
the filter, and also insures the highest oxidation of any arsenic or antimony present
in the filtrate. If five c.c. are insufficient to
impart a permanent tinge to the filtrate,
more must be added. Again wash the filter
and then boil the filtrate, which usually does
not exceed seventy-five c.c. in bulk, to
thoroughly expel the excess of bromine.
Remove from the heat and add ammonia
water in slight excess (ordinarily add five
c.c. of strong ammonia), and then acidify
with acetic acid. The addition of three c.c.
of the glacial acid is usually sufficient. A
large excess of acetic acid does no harm, but
is not necessary, except in the presence of
sufficient arsenic to cause a precipitate of

copper arseniate. This may require consid-
erable acetic acid for its solution. If not dissolved at this stage, it is taken up slowly
later on, and the titration may become very
tedious before the true end-point is finally
reached. Proceed with this acetic acid solution precisely as described in the standardization of the hyposulphite, and calculate the
percentage of copper from the amount of
hyposulphite required.

With high percentages it is advisable to
wash the precipitated copper by decantation
as follows: Transfer the liquid and copper
in the beaker to the original flask, and set
the beaker and aluminum aside temporarily.
Decant through the filter and wash the copper perhaps three times by decantation with
cold dilute ½ s. water, using about 20 c.c.
each time. Now place the flask under the
funnel, heat the five c.c. of acid in the beaker
as before, and pour it through the filter. Do
not wash for the moment, but remove the

flask, replacing it under the funnel with the beaker, and heat the acid until all the copper is dissolved. Now return the flask under the funnel and proceed with the washing. Boil off the red fumes and continue as described above.

Notes.—According to the equation previously given, half a gram of pure copper requires 2.62 grams of potassium iodide. While direct experiment shows this to be apparently true, yet with only the theoretical amount of iodide present the reaction is slow, and in fact does not appear to proceed to completion until during the titration, which is thereby unduly prolonged. It is best, therefore, to use not less than three grams in any case. An excess does no harm. Zinc and silver do not interfere. Lead and bismuth are without effect, except that by forming colored iodides they may mask the approach of the end-point before adding starch. Lead is practically removed as sul-

phate by the first filtration. If bismuth is suspected in appreciable amount, simply add the starch earlier in the titration. Arsenic and antimony when fully oxidized as described have no influence. The return of the blue tinge in the titrated liquid after long standing is of no significance, but a quick return which an additional drop or two of the hyposulphite does not permanently destroy is usually an evidence of faulty work.

A Modification of Low's Modified Cyanide Method.*

Bring the copper into solution in the same way as in the modified cyanide method (page 110). Evaporate the solution until all nitric acid is driven off and dense white fumes appear. Dilute with water until there is about five times as much water as sulphuric acid. Put 2 or 3 pieces of aluminum (25 gauge and about 40 mm. square, with one

---

* The Journal of the American Chemical Society, Vol. XXIV., page 478.

corner of each piece turned up) into the beaker containing the solution. Boil (about 5 minutes) until all the copper is precipitated on the aluminum.

Wash the precipitated copper into a tared Gooch crucible (use only sufficient asbestos fiber to make a good filter), wash, finally wash with alcohol (use only a little alcohol), burn off the alcohol, dry, and weigh as metallic copper.

The Cyanide Method as Modified by Thorn Smith.

Of the copper-bearing material take preferably such a weight that not over .25 gram of copper is present. Dissolve in a No. 2 beaker in 20 c.c. of nitric acid and the same amount of hydrochloric, if found necessary, and boil down to a low volume (almost dry). Then add from 7 to 10 c.c. of sulphuric acid and run down rapidly to white fumes. Cool and add 75 c.c. of water and two strips of sheet aluminum 2 inches long by one-half

wide and one-sixteenth thick, with opposite corners turned up to prevent sticking together. Boil until the blue color has entirely disappeared. The time required depends not so much on the amount of copper present as it does on the amount of ferric iron which must be reduced before the bulk of the copper will come down. It is well to continue the boiling for two or three minutes after the blue color is gone. At this point add 10 c.c. of hydrogen sulphide water and if more copper is precipitated as sulphide, boil until the copper sulphide settles or collects.

Filter by suction through a Gooch crucible in one piece, using a circular disk of filter-paper to collect the copper. If the solution filters slowly, owing to the presence of silica or other causes, the addition of a few drops of hydrofluoric acid will greatly hasten the matter. Do not add beyond a few drops, as too much is worse than none at all. Wash out the beaker with water con-

taining a little hydrogen sulphide, leaving the aluminum strips in the beaker. To the beaker add 5 c.c. of nitric acid by the aid of a pipette of that capacity, letting the acid run down the side of the beaker in such a way that all portions of the beaker will be touched and the adhering copper driven to the bottom. Heat to boiling. Remove the Gooch from the pump and invert over a clean No. 2 beaker, holding between the thumb and forefinger, and punch the paper disk out and into the beaker with a platinum poker. Now hold the Gooch right side up and still over the beaker, pour the nitric acid contained in the first beaker through the Gooch so that all portions of the interior will be touched and the copper thus washed through the Gooch. To the beaker just emptied add 10 cubic centimeters of strong bromine water and pour through the Gooch. Repeat this operation with bromine water, but reverse the Gooch and pour over the bottom.

The Gooch should be perfectly clean at the end of the operation. Cover the beaker and boil off the bromine. Then filter through an ordinary filter-paper, or, preferably, use the Gooch again. Wash well with hot water and add 10 c.c. of ammonia. Cool and titrate with a potassium cyanide solution equal to from .005 to .007 gram of copper per cubic centimeter. The best cyanide to use is the 98 per cent pure.

In titrating, the cyanide may be added quite rapidly at first, but as the color begins to fade the cyanide must be added much slower, and finally, near the end, several seconds must elapse between the drops. Some titrate to a faint pink and others to a colorless solution. The former is much the better practice. Whatever method is adopted should be rigidly adhered to. For standardizing, an ore of known copper content as determined by the electrolytic method is preferable to copper foil or any salt. In

sulphide ores the treatment with nitric acid
must be proceeded with slowly, as too much
heat at the beginning will prevent oxidation of sulphur.   At the end of the treatment with sulphuric acid the sulphur
remaining should appear as a yellow
globule.

In boiling with the strips of aluminum the
solution often darkens, especially near the end
of the precipitation, but this need cause no
alarm, as the addition of the hydrogen sulphide will clear it up.   The addition of hydrogen sulphide is necessary, as in most cases
the copper is not completely precipitated by
the aluminum.   The wash-water should contain a little hydrogen sulphide water, because
the moist precipitated copper oxidizes easily
and is apt to dissolve in the acid present.   It
must be remembered that the commercial
sheet aluminum or wire often contains appreciable traces of copper and that aluminum
is somewhat soluble in nitric acid, hence

should not remain in contact with it longer than necessary or aluminum hydroxide will appear in excess in the final operation. The cyanide method gives satisfactory results, but cannot be depended upon for accuracy.

CHAPTER XII.

Some assayers check their fire-assays of lead by a wet method. For the convenience of those who desire a rapid method to check their fire-assays, the ammonium molybdate* method is here given.

This method is based upon the fact that ammonium molybdate, when added to a hot solution of lead acetate, will give a precipitate of molybdate of lead ($PbMoO_4$), which is insoluble in acetic acid. Any excess of ammonium molybdate will give a yellow color with freshly prepared solution of tannin.

Indicator.—A freshly prepared solution of 1 part tannin in 300 parts water.

Standard Solution.—The standard solution

---

* H. H. Alexander, The Engineering and Mining Journal, April 1st, 1893.

† When elements are present that interfere with this method, consult a work on volumetric analysis. We cannot take the space in this book to describe long wet methods.

of ammonium molybdate is prepared by dissolving 9 grams of ammonium molybdate in
1000 c.c. of water. One c.c. of this solution
will equal about 0.01 gram lead. If the solution is not clear, it can be made so by adding
a few drops of ammonium hydrate.

Standardizing.—Weigh out 0.300 gram of
pure sulphate of lead, and dissolve it in hot
ammonium acetate; then acidify with acetic
acid, and dilute with hot water to 250 c.c.
Heat to boiling, and add from a burette the
molybdate solution, prepared as above mentioned, until all the lead is precipitated as a
white precipitate. This is ascertained by
placing drops of tannin solution upon a porcelain plate, and then to these drops is added
a drop of the solution tested, after each addition of ammonium molybdate. As long as
the lead is in excess, no coloration is produced; but as soon as the molybdate is in excess, a yellow color is produced with the tannin (0.300 gram $PbSO_4 \times 0.68317 = 0.20495$

gram Pb). The solution in the beaker should
be stirred after each addition of molybdate
solution before the drop-test is made. From
the number of molybdate solution used, the
value of one c.c. is calculated in the usual way.

Assay.—To determine the lead in ore or
other material, weigh out 0.5 or 1.0 gram of
the substance, according to the percentage of
lead. If the substance contains 30% or more
lead, 0.5 gram will be sufficient. Treat the
sample weighed out in a porcelain casserole
with 15 c.c. strong nitric acid and 10 c.c.
strong sulphuric acid. Heat until all the
nitric acid is expelled, which is indicated by
fumes of sulphuric anhydride coming off;
then allow it to cool, and dilute with cold
water; stir, then boil until all soluble sulphates are brought into solution. Now filter,
leaving as much of the precipitate in the
casserole as possible. Now wash twice with
hot dilute sulphuric acid and once with cold
water. The sulphate of lead remaining in the

casserole is next dissolved with hot ammonium acetate; pour the hot solution on the filter and allow it to run into a clean beaker. This operation is repeated until all the sulphate of lead is dissolved. Wash out the casserole thoroughly with hot water into the filter. Acidify the solution with acetic acid, dilute up to 250 c.c. with hot water. Now heat to boiling and run in from a graduated burette the standardized solution of ammonium molybdate until all the lead is precipitated, stirring the solution after each addition of molybdate, and testing a drop of the solution, after each addition of molybdate, on a porcelain plate with the tannin solution. From the number of c.c. of the molybdate solution used, calculate the per cent of lead.

The lead determination can easily be made in 30 minutes.

Arsenic, antimony and phosphorus do not interfere with this method, as they pass through the filter in solution.

APPENDIX A.

Cupels.—If the bone-ash used in making cupels is coarse, the cupels will be too porous, and much silver and some gold will be carried into the cupel. If the bone-ash is very fine and the cupels are compressed very hard, they will be too dense, may not absorb the litharge as fast as formed, and will crack in drying, and on becoming saturated with litharge. If the cupels are not compressed hard enough, they will be too porous; if compressed too hard, they will be too dense. The results in these cases are the same as stated above (see also p. 3). Some assayers take one part of wheat flour and mix it thor-

134

oughly with ten parts of bone-ash before moistening, and compress the cupels hard. On heating the cupels in the muffle before the buttons are dropped into them, the flour burns out, leaving the cupels porous.

From what has been stated, it is evident that cupels should be made of fine bone-ash, and of such hardness that they will absorb the litharge, and as little of the precious metals as possible. A cupel made of fine bone-ash, and of such hardness that, when dry, a fall of two to two and one-half feet will not break it, will absorb very little silver. By filling the cupel-mould about two-thirds full of coarse bone-ash, and the remainder, which forms the bowl of the cupel, with extra fine bone-ash, a cupel can be made that will absorb the litharge, and will absorb very little silver.

Cupels will be stronger and less liable to crack in drying, if a strong solution of sodium or potassium carbonate is used instead of

water to moisten the bone-ash. The bone-ash should not be sufficiently moistened to feel wet (see p. 1).

Cupels should be dried slowly before they are used. If they are put into the hot muffle while moist, they will fall to pieces. If buttons are dropped into the cupels before all the moisture and the gases are expelled, loss may result from spirting.

Sampling.—Great care should be taken in taking an average sample of ore. Unless the sampling is properly done, the results of the assay are worthle s. Not only should great care be taken in taking an average sample across the vein of ore in the mine at different places, but also in sampling the ore so taken for an assay sample (see p. 29).

Fluxing and Fusion.—In the chapter on slags, it has been explained that assay charges are made up of basic and acid substances in such proportions that they will combine and form liquid slags. The general reactions are

indicated, but, as stated on page 49, when so many substances are thrown together, many other reactions take place.

The gangue must be so fluxed that all becomes fusible, or the precious metals cannot be released, and collected in a button of lead in the bottom of the crucible. For example, if ferric oxide, Fe_2O_3, forms part of the gangue materials, the charge must contain sufficient reducing agents to reduce the ferric oxide to ferrous oxide, FeO. The ferric oxide would not fuse, but would be entangled or absorbed by the soda or slag, and the gold and silver contained in the ferric oxide would be carried into the slag with it; but, if the ferric oxide is reduced to ferrous oxide, the ferrous oxide unites with the silica, etc., and the compound so formed becomes fusible, and allows the precious metals to drop to the bottom of the crucible by virtue of their greater specific gravities (see also under Corrected Assays).

Cupellation.—In the operation called cupellation, the lead is oxidized forming litharge. The other base metals also oxidize. Litharge is fusible at a bright red heat, and, when fused, has the property of dissolving or absorbing oxides of other metals. Bone-ash has the property of absorbing melted litharge, and a certain amount of other oxides that may be dissolved by the litharge. Gold, silver, platinum, and some of the rarer metals, do not oxidize by this operation, and, therefore, are not dissolved or absorbed by the litharge, and hence not carried by it into the bone-ash.

When a large amount of base metals is present, the litharge formed may not be sufficient to carry all the oxides into the cupel. If this is the case, some of the oxides remain on the cupel, known as a scoria. An addition of lead might have carried all the oxides into the cupel. A scoria also forms on the cupel if the lead button is not entirely freed

from slag. The scoria may have entangled some lead which carried the gold and silver. Hence the scoria may contain some of the precious metals, which can be recovered by crucible or scorification assay.

After charging in the buttons, the muffle should be closed until they uncover. The muffle should be hot enough to start a rapid oxidation. Then the temperature should be lowered to form litharge crystals, and the buttons should be allowed to blick at a higher temperature.

The muffle can be cooled by checking the fire, or by putting cold scorifiers, crucibles, or cupels into the part of the muffle that is too hot, replacing them when hot, until the desired temperature is secured.

The assayer should aim to approach these somewhat ideal conditions:

1. To flux the gangue so that every particle of it becomes fusible and forms a liquid slag.

2. To reduce the least amount of lead that will collect all the gold and silver.

3. To have a cupel that is fine and hard enough to absorb all the litharge, and none of the gold and silver.

4. To oxidize the button in the least time possible, avoiding bad effects of too rapid oxidation.

5. To have the muffle hot enough to start a rapid oxidation, then cool sufficiently to form litharge crystals, and allow the buttons to blick at a higher temperature.

APPENDIX B.

ATOMIC WEIGHTS AND TABLES.

Atomic Weights.

(From the Ninth Annual Report of Committee on Atomic Weights,
Journal American Chemical Society, XXIV. No. 3)

Element.	H = 1.	O = 16	Element.	H = 1.	O = 16.
Aluminum.	26.9	27.1	Neodymium. . .	142.5	143.6
Antimony.	119.5	120.4	Neon.	19.8	19.94
Argon.	39.6	39.96	Nickel.	58.25	58.70
Arsenic.	74.45	75.0	Nitrogen.	13.93	14.04
Barium.	136 4	137.40	Osmium.	189.6	191.0
Bismuth.	206.5	208.1	Oxygen.	15.88	16.000
Boron.	10.9	11.0	Palladium.	106.2	107.0
Bromine.	79.35	79.95	Phosphorus. .	30.75	31.0
Cadmuim.	111.55	112.4	Platinum.	193.4	194.9
Cæsium.	131.9	132.9	Potassium , . . .	38.82	39.11
Calcium.	39.8	40.1	Praseodymium	139.4	140.5
Carbon.	11.9	12.0	Rhodium.	102.2	103.0
Cerium.	138.0	139.0	Rubidium.	84.75	85.4
Chlorine.	35.18	35.45	Ruthenium. . . .	100.9	101.7
Chromium.	51.7	52.1	Samarium.	149.2 ?	150.3 ?
Cobalt.	58.55	59.0	Scandium.	43.8	44.1
Columbium. . . .	93.0	93.7	Selenium.	78.6	79.2
Copper	63.1	63.60	Silicon.	28.2	28.4
Erbium.	164.7	166.0	Silver	107.11	107 92
Fluorine.	18.9	19.05	Sodium	22.88	23 05
Gadolinium. . . .	155.2	156.4	Strontium.	86.95	87.60
Gallium.	69.5	70.0	Sulphur.	31.83	32.07
Germanium. . . .	71.9	72.5	Tantalum.	181.5	182.8
Glucinum.	9.0	9.1	Tellurium. , . . .	126.1	127.7
Gold.	195.7	197 2	Terbium.	158.8	160.
Helium.	3.93	3.96	Thallium	202.61	204.15
Hydrogen.	1.000	1.008	Thorium.	230.8 ?	232.6 ?
Indium	131.1	114.0	Thulium	169.4	170.7
Iodine.	125.89	126.85	Tin. . , ...	118.1	119.0
Iridium.	19 .7	193.1	Titanium	47.8	48.15
Iron.	55.5	55.9	Tungsten.	182.6	184.
Krypton.	81.15	81.76	Uranium.	237.8	239.6
Lanthanum. . . .	137.6	138.6	Vanadium . . .	51.0	51.4
Lead.	205.36	206.92	Xenon.	127.	128.0
Lithium.	6.97	7.03	Ytterbium. . . .	171.9	173.2
Magnesium. . . .	24.1	24.3	Yttrium	88.3	89.0
Manganese.	54.6	55.0	Zinc.	64.9	65.4
Mercury.	198.50	200.0	Zirconium.	89.7	90.4
Molybdenum. .	95.3	96.0			

Troy Weight.

```
 24 grains =    1 dwt.
 480    "    =  20  "  = 1 oz.
5760    "    = 240  "  = 12  "  = 1 lb. = 22.816 cu. in. of
                distilled water at 62° Fahr.
```

Avoirdupois Weight.

```
    1 drachm =                27.34375 grains Troy.
   16    "    =      1 oz.= 437.5        "      "
  256    "    =     16  " =    1 lb.=  1.2153 lb. Troy.
 6400    "    =    400  " =   25  " =  1 quarter.
25600    "    =   1600  " =  100  " =   4    "
512000   "    = 32000  " = 2000  " = 80    "
                               = 20 cwt. = 1 ton.
```

Relations between Various Weights.

	Grams.	Troy Grs.	Troy Ozs.	Troy Lbs.
1 Milligram =	.001=	.01543		
1 Centigram =	.01 =	.15432		
1 Decigram =	.1 =	1.5432		
1 Gram =	1.	= 15.432 =	.032 =	.00267
1 Decagram =	10.	=	.321 =	.02679
1 Hectogram=	100.	=	3.215 =	.26792
1 Kilogram =	1000.	=	32.150 =	2.6792
1 Myriagram=	10000.	=	26.792	
1 Quintal =	100000.	=	267.92	
1 Tonneau =	1000000.	= 2679.2		

Relations between Various Weights—(*Continued*).

		Avoir Ozs.		Avoir Lbs.
1 Gram	=	.03528	=	.0022047
1 Decagram	=	.3528	=	.022056
1 Hectogram	=	3.52758	=	.22046
1 Kilogram	=	35.2758	=	2.2046
1 Myriagram	=........................			22.046
1 Quintal	=.......................			220.46
1 Tonneau	=.........			2204.6

.064798 gram = 1 Troy grain.

1 gram = 15.43235 Troy grains.

1 A. T. = 29.166 grams = 450.0999 Troy grains.

To Convert Centigrade and Fahrenheit Degrees.

$$9/5\,C° + 32 = F° ;$$
$$5/9(F° - 32) = C°.$$

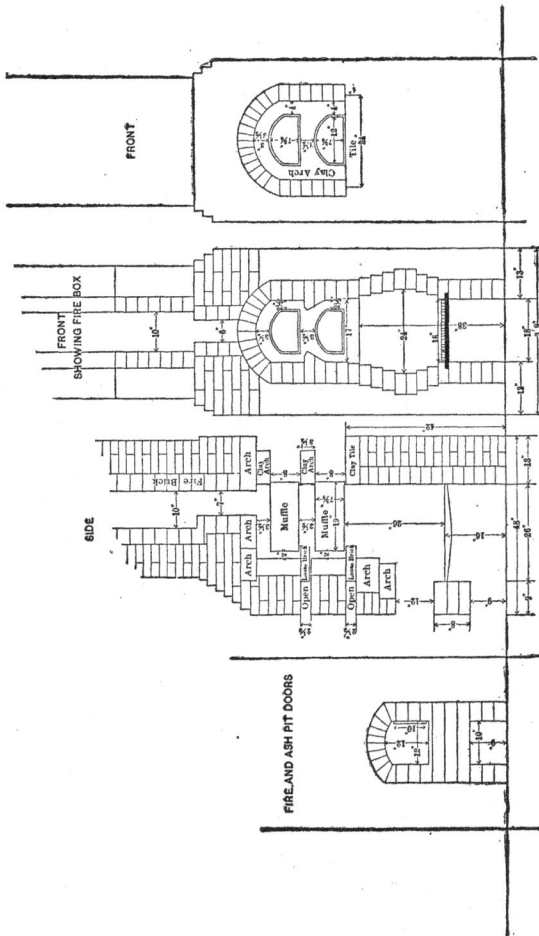

WOOD-BURNING MUFFLE-FURNACE.

(From Eng. and Mining Journal, Dec, 20, 1902.)

INDEX.

A

	PAGE
Amalgamation Test	82
Ammonium Carbonate, in roasting	70
Ammonium Molybdate Method for Lead	130
Assay of Gold and Silver Ores	9, 24, 61
Corrected Assay	70
Notes on Assay of Ores	29, 134–140
Assay of Copper Ores for Gold and Silver	91–99
Ores containing Coarse Metals	88
Copper Materials for Gold and Silver	91–99
Copper	108
Cupels	71
Lead	26, 44, 130
Mercury	100
Slags	71
Preliminary Assay	63
To make the Assay	9
To prepare the Assay	5
Assay-ton Weights	8

B

| Bicarbonate of Soda | 48 |
| Borax | 51 |

C

PAGE

Charcoal... 54
Chlorination Test..................................... 86
Copper, Determination of............................ 108
    Cyanide Assay................................. 108
    Iodide Assay.................................. 114
    Ores, Assay for Gold and Silver............... 91–99
    Materials, Assay for Gold and Silver........... 91–99
Crucible Charges............ 10, 58, 72, 73, 74, 76, 77, 78
Cupels.. 1, 134
    Assay of...................................... 71
Cupellation................................... 18, 34, 138
    Influence of Base Metals on................... 81
    Notes on.................................. 34, 138

D

Desulphurization by Means of Iron Nails............. 65b
Determination of Copper............................ 108
    Gold and Silver......................... 10, 24, 61
        in Copper Ore........................ 91–99
        in Copper Matte...................... 91–99
        in Copper............................ 91–99
        in Cupels............................ 71
        in Slags............................. 71

F

Fire-assay for Gold and Silver Ores (Crucible). 10, 24, 29, 61
    (Scorification)............................. 24, 43
Flour... 52
Fluxes.............................. 3, 4, 46, 47, 76–80
Fusion in Crucible................................. 13
    Notes on.................................. 32, 136

G

PAGE

Godshall's Method................................. 96*a*

I

Iron, Flux..................................... 54, 104
in Assays.................................... 65*b*

L

Lead, Determination by Fire-assay................ 26, 44
by the Wet Method........................ 130
Metallic...................................... 54*a*
Notes on the Fire-assay.................... 44, 140
Fluxes...................................... 26, 46
Litharge... 50

M

Mercury, Assay of................................ 100

P

Parting... 21
Notes on..................................... 37
Potassium Cyanide................................ 53
Potassium Nitrate................................ 54
Preliminary Assay.............................. 63, 64

R

Retorting... 85
Roasting.. 67
with Ammonium Carbonate..................... 70

S

PAGE

Salt.. 54*a*
Samples and Sampling............................ 29, 136
Scorification Assay................................. 24
    Notes on...................................... 43
Scorification Charges............................. 79, 80
Silica.. 52
Slags.. 55
    Assay of...................................... 71
Sodium Bicarbonate................................ 48
Special Methods.................................... 74

T

Tables.. 141
    Atomic Weights............................. 141
    Avoirdupois Weight.......................... 142
    Metric, or French Weights.................... 142
    Troy Weight................................. 142
Telluride Ores....................................... 74
Test-lead... 54*a*

V

Van Liew's Method................................ 96*b*

W

Whitehead's Method............................... 96

Disclaimer ... Please Read

ASSAYING MANUAL

THE FIRE ASSAY OF GOLD, SILVER, AND LEAD, INCLUDING AMALGAMATION AND CHLORINATION TESTS

BY
ALFRED STANLEY MILLER

Professor of Mining, Metallurgy, and Geology
University of Idaho

THIRD EDITION

REVISED AND ENLARGED

2008

Wexford